On the Perception of Self-motion: from Perceptual Laws to Car Driving Simulation

Dissertation

zur Erlangung des Grades eines
Doktors der Naturwissenschaften

der Mathematisch-Naturwissenschaftlichen Fakultät
und der Medizinischen Fakultät
der Eberhard-Karls-Universität Tübingen

vorgelegt von

Alessandro Nesti
aus Varese, Italy

Oktober 2014

Bibliografische Information der Deutschen Nationalbibliothek

Die Deutsche Nationalbibliothek verzeichnet diese Publikation in der
Deutschen Nationalbibliografie; detaillierte bibliografische Daten sind
im Internet über http://dnb.d-nb.de abrufbar.

ISBN 978-3-8325-4013-5

Logos Verlag Berlin GmbH
Comeniushof, Gubener Str. 47,
10243 Berlin
Tel.: +49 (0)30 42 85 10 90
Fax: +49 (0)30 42 85 10 92
INTERNET: http://www.logos-verlag.de

Tag der mündlichen Prüfung:	27 April 2015
Dekan der Math.-Nat. Fakultät:	Prof. Dr. W. Rosenstiel
Dekan der Medizinischen Fakultät:	Prof. Dr. I. B. Autenrieth
1. Berichterstatter:	Prof. Dr. Heinrich H. Bülthoff
2. Berichterstatter:	Prof. Dr. Uwe J. Ilg
Prüfungskommission:	Prof. Dr. Heinrich H. Bülthoff
	Prof. Dr. Uwe Ilg
	Prof. Dr. Hanspeter A. Mallot
	Prof. Dr. Stefano Ramat

I hereby declare that I have produced the work entitled: "On the Perception of Self-motion: from Perceptual Laws to Car Driving Simulation", submitted for the award of a doctorate, on my own (without external help), have used only the sources and aids indicated and have marked passages included from other works, whether verbatim or in content, as such. I swear upon oath that these statements are true and that I have not concealed anything. I am aware that making a false declaration under oath is punishable by a term of imprisonment of up to three years or by a fine.

Tübingen, 29 September 2014

Alessandro Nesti

ACKNOWLEDGEMENTS

This thesis concludes an adventure about science and friendship. Here, I want to thank all the persons who, in one way or another, contributed to the success of this work, turning these years into a wonderful experience that I will cherish for the rest of my life.

I want to express my greatest gratitude to Professor Heinrich Bülthoff for his support of my ideas and my projects and for the opportunity to work in his group. I cannot imagine a more dynamic and stimulating environment for conducting scientific research. I also want to thank Paolo Pretto, Karl Beykirch, Michael Barnett-Cowan and Paul MacNeilage for their close supervision, for their valid collaboration and for their friendship.

I am thankful to the Professors Uwe Ilg and Stefano Ramat for being part of my advisory board and for providing insightful comments on my work. I am particularly grateful to Professor Uwe Ilg for reviewing my thesis and to Professor Stefano Ramat for encouraging me to pursue the PhD abroad, a decision that I will never regret.

I thank all my colleagues and collaborators within and outside the institute for making these years in Tübingen so interesting and full of enjoyable moments. A special thank goes to Florian Soyka, a precious resource for comments and advices and a great office mate. Furthermore, I want to thank the administrative and technical staff of the Max Planck and of the Graduate School for being so efficient and helpful. In particular, thanks to Maria Lächele for her assistance in setting up the experiments with the CyberMotion Simulator.

Thanks to all those people who, outside the institute wall, immensely enriched my years in Tübingen. My flatmates, the Tübingen jugglers, the various beach volleyball players and my two volleyball teams SSC and TSG, thanks for all the fun we had together.

Most of all, I want to thank Mariangela for sharing every moment and every emotion. Thanks for being strong when I needed to be strong, for being patient, and encouraging.

And lastly, a very important thank to my family: my sister Francesca and my parents Lucia and Antonio for their lifelong love and support.

ABSTRACT

Everyday life requires humans to move through the environment, while completing crucial tasks such as retrieving nourishment, avoiding perils or controlling motor vehicles. Success in these tasks largely relies in a correct perception of self-motion, i.e. the continuous estimation of one's body position and its derivatives with respect to the world. The processes underlying self-motion perception have fascinated neuroscientists for more than a century and large bodies of neural, behavioural and physiological studies have been conducted to discover how the central nervous system integrates available sensory information to create an internal representation of the physical motion. The goal of this PhD thesis is to extend current knowledge on self-motion perception by focusing on conditions that closely resemble typical aspects of everyday life.

In the works conducted within this thesis, I isolate different components typical of everyday life motion and employ psychophysical methodologies to systematically investigate their effect on human self-motion sensitivity. Particular attention is dedicated to the human ability to discriminate between motions of different intensity. How this is achieved has been a fundamental question in the study of perception since the seminal works of Weber and Fechner. When tested over wide ranges of rotations and translations, participants' sensitivity (i.e. their ability to detect motion changes) is found to decrease with increasing motion intensities, revealing a nonlinearity in the perception of self-motion that is not present at the level of ocular reflexes or in neural responses of sensory afferents. The relationship between the stimulus intensity and the smallest intensity change perceivable by the participants can be mathematically described by a power law, regardless on the sensory modality investigated (visual or inertial) and on whether visual and inertial cues were presented alone or congruently combined, such as during natural movements. Individual perceptual law parameters were fit based on experimental data for upward and downward translations and yaw rotations based on visual-only, inertial-only and combined visual-inertial motion cues. Besides wide ranges of motion intensities, everyday life scenarios also provide complex motion patterns involving combinations of rotational and translational motion, visual and inertial sensory cues and physical and mental workload. The question of how different combinations of these factors affect motion sensitivity was experimentally addressed within the framework of driving simulation and revealed that sensitivity might strongly decrease in more realistic conditions, where participants do not only focus on perceiving a "simple" motion stimulus (e.g. a sinusoidal profile at a specific frequency) but are, instead, actively engaged in a dynamic driving simulation.

Applied benefits of the present thesis include advances in the field of vehicle motion simulation, where knowledge on human self-motion perception supports the development of state-of-the-art algorithms to control simulator motion. This allows for reproducing, within a safe and controlled environment, driving or flying experiences that are perceptually realistic to the user. Furthermore, the present work will guide future research into the neural basis of perception and action.

CONTENTS

Cover Picture: Max Planck Institute for Biological Cybernetics

1

INTRODUCTION

1.1 SELF-MOTION PERCEPTION IN EVERYDAY LIFE

While standing on the bus on the way home, how do we realize when the driver initiates a sudden brake? While our body dangerously leans forward, multiple sensory cues are available for detecting the balance-threat. A first clue would certainly be the feeling from our whole body as it starts to decelerate. Additionally, we would realize a change in the relative position of the upper body with respect to the legs and feet, and we would see the world image flowing outside the window slowing down to a full stop. Possibly, we might even hear a fellow passenger gasping from the surprise. A prompt interpretation of these cues gives us the rapid awareness of the situation that we need in order to firmly hold to the bus handle and prevent an embarrassing and potentially dangerous fall. The described scenario is just an example of the countless everyday life situations where the central nervous system (CNS) is called upon to create an internal representation of our own physical motion through the environment. This process of estimating the state of the body (i.e. its position, orientation and their derivatives) with respect to the world is generally referred to as self-motion perception[1]. It is easy to imagine how self-motion perception plays

---

[1] Sometimes a distinction is made between movements of parts of the body of the observers (self-motion) and movements of the whole body of the observer (ego-motion).

vital evolutionary roles for the survival of living organisms, increasing the probability of successfully completing actions such as retrieving nourishment or avoiding perils.

The large body of established research in basic self-motion perception is unsurprising, considering how frequently it factors in our daily activities. Research on self-motion perception is further supported by a variety of applied fields where immediate and concrete benefits directly follow from the results of a set of experiments. The work contained in this thesis is sensitive to the quest of robustly supporting practical goals, and does so with basic and systematic studies to investigate the processes that underlie self-motion perception. Experiments were, therefore, designed to provide insight into the influence of sensory and cognitive factors on self-motion perception while at the same time being of direct interest for the field of vehicle motion simulation. This is further discussed in sections 1.4, 1.5 and in chapter 6. In the remainder of this section, some of the main applied fields related to the study of self-motion perception are presented and briefly discussed.

Perception-based motion simulation

Self-motion is sometimes experienced in potentially dangerous scenarios such as when learning to drive in the city traffic or flying an aircraft. One way to improve safety and reduce the number of accidents is to allow drivers (or pilots) to repeatedly experience the situation of interest inside a dynamic motion simulator, where their errors have harmless consequences. Dynamic motion simulators generally consist of a moving base where the driver sits, a visual display, audio speakers and control devices such as a steering wheel and pedals. A virtual model of the simulated vehicle processes the control inputs from the driver and calculates how the real vehicle would respond to the motion command in the specific situation. In spite of technological advances, faithful reproduction of the simulated manoeuvres will always be constrained by the physical limits of the motion system, particularly when those manoeuvres involve high or sustained accelerations. It is therefore a challenging problem to select feasible motions that prevent drivers from perceiving a conflict between expected and experienced motion. This is commonly done by implementing knowledge about human self-motion perception in the control framework of motion based simulators (Nahon and Reid 1990; Telban et al. 2005). Commonly employed "perceptual tricks" include using movements, whose intensities are so low that they cannot be perceived by humans. This allows for instance to reposition the simulator every time one of its actuators approaches the end of its operational range (motion washout). Moreover, unperceivable movements are sometime employed to tilt the driver without him/her noticing the tilt: the resulting sensation of being forced into (or away from) the seat, combined with the appropriate visual stimulus, evokes the illusory perception of a sustained linear acceleration. State-of-the-art motion algorithms dedicate increasing attention to the research field of self-motion perception and promptly implement novel findings, such as those presented in this thesis, to obtain higher simulation realism (see Garrett and Best (2010) for a review on motion cueing algorithms). Perception-based motion simulation constitutes an important part of this thesis and will be further discussed in greater details (see sections 1.4.3 and 1.5 and chapter 6).

Diagnosis and rehabilitation of balance disorders

Balance disorders are very common especially in the elderly population (Wall et al. 2003; Wall and Weinberg 2003). The causes often rely in injuries or diseases that result in to a deficiency of either the peripheral self-motion sensory organs or the CNS components that process self-motion information. Symptoms include dizziness, vertigo, nausea and an impaired ability to walk and maintain balance. Because eye movements, like self-motion perception, receive inputs from the vestibular system, common diagnosis protocols for balance disorders rely on eye movements analysis in response to passive or active head movements (Bárány 1921; Halmagyi and Curthoys 1988). Eye movements are relatively easy to observe and measure. Their abnormalities indicate possible deficiencies of the vestibular system or the neural path connecting the vestibular system to the eyes. Nonetheless, perception and such reflexes do not always correlate (Kanayama et al. 1995; Merfeld et al. 2005a, b). Therefore, quantitative measures of self-motion perception such as perceptual thresholds (described in section 1.3.1) are a helpful tool that can be used in combination to localize the disorder source (Merfeld et al. 2010; Agrawal et al. 2013). Furthermore, they could help in assessing the efficacy of specific rehabilitation protocols and quantitatively evaluate the benefits of balance prostheses, such as vibrotactile vests, which signal the occurrence of a fall. Another interesting prosthesis (Wall et al. 2003; Wall and Weinberg 2003) that recently drew attention, and could benefit from quantitative perceptual evaluation, relies on galvanic vestibular stimulation (GVS) of the afferent fibres. Using cutaneous or implanted electrodes, GVS prostheses deliver a small current calibrated to compensate for a deficient vestibular system and evoke the appropriate self-motion perception and balance reflexes.

Motion sickness

Motion Sickness is a common "disorder" that can be associated with self-motion perception. It is commonly experienced when traveling in a vehicle (e.g. a ship) or when exploring virtual environments and gives rise to symptoms such as dizziness, fatigue or nausea. Underlying the most widely accepted motion sickness theories and models (Reason 1969; Oman 1982) is the sensory conflict hypothesis (Beadnell 1924). It states that motion sickness arises whenever visual and inertial sensory cues deviate from normal daily patterns and conflict with each other. For example, rocking on a ship with no windows is a typical scenario that induces motion sickness. This is because the ship motion results in inertial cues that are inconsistent with the visual scene, which, in this case, is stable with respect to the ship and not with respect to the real world. Motion sickness recently was entitled with further attention due to the advances and increased popularity of virtual environments that humans can explore while remaining physically stationary (e.g. by means of head-mounted displays). Motion sickness research currently focuses on understanding why some visual-inertial conflicts trigger motion sickness whereas others lead instead to a reinterpretation of the perceived self-motion (Lackner 2014). This could results in effective prevention strategies.

Study and prevention of human errors in vehicle control

Self-motion perception does not always succeed in estimating the true physical motion of the body for a variety of different reasons, which relate to (but are not limited to) sensors dynamics, neural variability and multisensory integration. Specific examples of illusory or misinterpreted self-motion are presented and discussed on several occasions throughout this thesis. It is easy to imagine how perceiving a motion that is not there, as well as not perceiving an occurring motion, might lead to dangerous behaviours. For instance, an aircraft pilot flying through a cloud might misinterpret a forward acceleration for an upward tilt of the cockpit and, in an attempt to react to it, enter a stall position and crash into the ground (Newman 2012). In an attempt to predict the occurrence of erroneous self-motion perception several computational models have been proposed (Zaichik et al. 1999; Bos and Bles 2002; Zupan et al. 2002; Newman et al. 2012) and are constantly improved to account for the latest experimental findings. Matching the physical aircraft motion as recorded by motion sensors with the output of a pilot perception model is a valuable way of predicting perceptual instances of spatial disorientation. This allows for determining whether pilot spatial disorientation played a role in a given accident and to identify disorienting physical motion profiles that can then be safely experienced by pilots during training in flight simulators.

Teleoperation of mobile robots

Unmanned aerial vehicle (UAV), also known as drones, are nowadays widely employed in civil applications, for instance to explore and perform actions in dangerous environments or for surveillance. Although much of the control is automated, a human operator remains an essential element in the control of UAV due to his/her degree of flexibility and problem solving skills. An active research field is the study of useful motion feedback to increase teleoperators' awareness of the UAV status. This avoids overloading the visual sensory channel with positional and inertial sensory information, in addition to the live image stream from the UAV's on-board camera, and can result in improved control performances (Robuffo Giordano et al. 2010a; Lächele et al. 2014).

1.2 HUMAN RESOURCES FOR SELF-MOTION PERCEPTION

Compared to many physical stimuli such as colours, sounds or heat, self-motion stimulates a surprisingly large number of human sensory organs. The inertial forces generated in response to whole-body acceleration stimulate the vestibular organs as well as different sensory receptors of the somatosensory system. Moreover the eyes provide the CNS with visual images "flowing" in the opposite direction of the head's motion. Whereas these are the main sensory systems involved in perceiving self-motion, additional information might be transmitted via other sensory modalities. For instance, auditory cues might inform about relative changes in distance and direction of a sound source with respect to the ears, a possible consequence of self-motion.

These section focuses on the description of the visual, vestibular and somatosensory systems as these are the sensory systems that are primarily involved in the experiments conducted in the present thesis work. Moreover, studies focussing on the contribution of cognitive rather than sensory factors to self-motion perception are reviewed and discussed here.

1.2.1 THE VISUAL SYSTEM

Sensory receptors in the eyes respond to light stimuli (Kandel et al. 2000). Although light stimuli do not inform about self-motion per se, the continuous variations of the retinal image (optic flow) that are experienced when moving through the environment provide important cues about the direction and intensity of linear and angular head movements (von Helmholtz 1925). Even in absence of inertial cues, optic flow alone can evoke balance responses (Van Asten et al. 1988; Wei et al. 2010) as well as a compelling illusion of self-motion, defined in the literature by the term *"vection"* (von Helmholtz 1925). A common instance of vection occurs when we sit in a stationary train and experience self-motion upon seeing neighbouring trains move. This illusion is widely exploited in the field of virtual reality to simulate self-motion without using motion simulators. In everyday life, however, the visual motion evoked by self-motion is most of the time congruent with the inertial stimulus, i.e. it is equal in intensity and flows in the opposite direction.

Besides optic flow, priors that are relevant for self-motion perception can be retrieved from the visual scene (Asch and Witkin 1948a; Asch and Witkin 1948b). Many natural scenes contain objects with a preferred orientation; for instance a tree usually grows vertical and points toward the sky, whereas the flat earth horizon is perpendicular to the direction of gravity. These visual elements, if present in the scene, intrinsically inform about head tilt with respect to an upright posture and contribute to a continuous estimate of head orientation in space.

It should be noted that the visual system, unlike the vestibular system (see section 1.2.2), does not respond exclusively to self-motion. Visual motion could also result from objects moving with respect to the head. It is generally believed that, when processing visual information, the CNS is able to appropriately attribute optic flow to self-motion or object-motion (Dichgans and Brandt 1978; Wertheim 1994). For example, a visual rotation of the surrounding environment is usually perceived as *object* motion immediately after the visual motion onset. Later, a perception of *self-*

motion builds up over time until the observer confidently believe that the visual scene is stationary and all the visual motion is perceived as self-motion. The well-established theory from Dichgans and Brandt (1978) argues for a reciprocal relationship between the perception of object motion and self-motion, although other models have been suggested (Wertheim 1994).

1.2.2 THE VESTIBULAR SYSTEM

The main peripheral components of the vestibular system (Kandel et al. 2000; Highstein et al. 2004) are 2 bony structures, the vestibular labyrinths, located in the inner ears (see Figure 1). Each labyrinth is composed by two otolith organs, the utricule and the saccule, and 3 semicircular canals.

The utricule and the saccule are sensitive to accelerations acting on the head. They are sack-like bony structures about 3 mm in the longest dimension which contain sensory hair cells (approximately 30.000 in the utricule and 16.000 in the saccule) arranged on a planar sensory epithelium, the macula. When the head is upright, the utricular and the saccular macula are approximately parallel and perpendicular to the ground, respectively. Within the macula, sensory cells are immersed in the otolithic membrane, a gelatinous membrane composed by a fluid, the endolymph, which contains several crystals of calcium carbonate, the otoconia (0.5-10 μm long). In response to head accelerations, displacements of the otolithic membrane will cause the hairs of the hair cells to bend, leading to an increase or decrease of the number of neural impulses traveling along the cells afferent fibres. Although each cell is sensitive to one specific acceleration direction, hair cells are oriented differently inside the macula, thus the combining neural signals from the utricule and the saccule allows to compute the 3d vector of head acceleration. However, in a gravitational environment, such as the one we inhabit, this vector (called gravito-inertial vector) is the resultant of both the gravitational acceleration and the inertial linear acceleration associated to self-motion. Indeed, both linear inertial accelerations and a gravitational

8

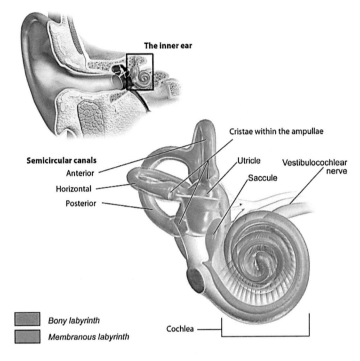

Figure 1 *In mammals, the inner ear is composed by a bony labyrinth containing the cochlea, which transduces sounds, and the vestibular system, which transduces inertial movements of the head. The membranous labyrinth runs inside the bony labyrinth and houses the sensory receptors. Sensory information is then transmitted, in the form of neural signals, to the CNS via the vestibulocochlear nerve, consisting of the vestibular nerve and the cochlear nerve. Adapted from Blausen[2].*

---

[2] "Blausen 0329 EarAnatomy InternalEar" by BruceBlaus - Own work. Licensed under Creative Commons Attribution 3.0 via Wikimedia Commons
http://commons.wikimedia.org/wiki/File:Blausen_0329_EarAnatomy_InternalEar.png# mediaviewer/File:Blausen_0329_EarAnatomy_InternalEar.png

The 3 semicircular canals (approximately 8 mm diameter) are sensitive to angular accelerations of the head. They are closed tubes filled with endolymph fluid and oriented perpendicularly to each other, with the horizontal canal tilted backward by approximately 30 degrees with respect to the ground when the head is upright. The other 2 canals, anterior and posterior, are oriented vertically, with the anterior canal plane from one ear parallel to the posterior canal plane of the opposite ear. Each semicircular canal has, at its end, a dilatation, the ampulla. There the fluid flow is interrupted by a gelatinous membrane, the cupula, an elastic swing-door-like structure that contains the sensory epithelium (crista ampullaris). Sensory cells of the crista (approximately 7.000), apart from having longer hairs and being all polarized in the same direction, are similar to those of the macula in the otoliths. In response to angular acceleration of the canals the cupula, rigidly attached to the canals, is bent by the endolymph, which lags behind due to its inertia. This causes the hairs of the sensory cells to bend and, in turn, modulates the firing rate of the afferent neurons. Note that, although the endolymph motion is generated by angular acceleration of the canals, for frequencies higher than ~0.1 Hz the hair cells deflection is proportional to angular velocity. This is due to the small internal diameter of the canals (approximately 0.3 mm) which increases the viscous properties of the endolymph and results in a mathematical integration of the input signal.

Besides informing the CNS about self-motion and body orientation, the vestibular system also controls postural and ocular reflexes and it is often referred to as the "balance organ". This definition is slightly misleading as it hinders the important contribution of the other before-mentioned human sensors. Indeed, most cases of vestibular disorders, quite common especially in the elderly population, only affects daily activities in the first months after the injury. In this regard it is extremely interesting to read Crawford's (1964) description of his personal experience of a vestibular nerve damage and how he learned in little time to regain the ability to walk, swim and play tennis.

1.2.3 THE SOMATOSENSORY SYSTEM

Besides visual and vestibular signals, the CNS can rely on somatosensory information to estimate the inertial acceleration of the body. This information arises from mechanoreceptors in the skin and from proprioceptors located in muscles and joints. It was further suggested by Mittelstaedt (1996) that an additional source of information comes from gravitoreceptors located along the trunk (possibly in the kidneys) that specifically respond to constant accelerations such as gravity.

Mechanoreceptors in the skin are sensory receptors that respond to mechanical pressure or distortion such as those generated by touching objects or stretching the skin (Kandel et al. 2000). These sensors are particularly involved during self-motion of a seated observer, such as a car driver. For instance, the back of the seat presses against the driver during forward acceleration and activates mechanoreceptors on the driver's back, which results in the sensation of being forced into the seat.

Proprioceptors provide information about the relative position of body segments by sensing muscle stretch and muscle tension. Particularly relevant for perceiving self-motion and maintaining balance is proprioceptive information from the neck muscles (Cohen 1961). This informs the CNS about body position relative to the head. Two important types of proprioceptors are the muscle spindle and the Golgi tendon organs (Kandel et al. 2000). Muscle spindles are located throughout the body of a muscle and their afferent signal is proportional to muscle stretch and stretch speed. Because active movements are generated through muscle contraction, stretch must be due to external forces acting on the muscle. Golgi tendon organs are located in the tendons that connect the bone with the muscle fibers of voluntarily controlled muscles. The afferent neural response is proportional to muscle tension, which usually increase as the result of a motor command. Interestingly, they do not respond well to stretch as most of the force is absorbed by the muscle itself during stretch.

Vestibular and somatosensory neural information are integrated very early, already in the vestibular nuclei (Cullen 2012). Moreover, self-motion stimuli, such as those employed throughout the present thesis work, as well as natural movements usually evoke both vestibular and somatosensory responses. Therefore, it seems unlikely that a distinction can be made between somatosensory and vestibular self-motion perception for these stimuli. Throughout the thesis, the term "inertial stimuli" is used to indicate those motion stimuli that physically accelerate the participants' body, stimulating both vestibular and somatosensory receptors.

1.2.4 COGNITIVE CONTRIBUTIONS

Sensory information is not the only resource available to the CNS: self-motion perception also relies on cognitive factors such as expectation of the motion that might (or is about to) occur. For instance, it has been shown that reports of illusory tilt perception during low-frequency lateral accelerations are cognitively reduced (if not suppressed) by prior knowledge about simulator capabilities (Wertheim et al. 2001). Expectation has also been shown to play a role in vection, facilitating its occurrence (i.e. lower onset latencies) when participants are sitting on a movable chair as compared to a fixed one (Palmisano and Chan 2004). The role of attentional factors was demonstrated by Hosman and Van der Vaart (1978): pilots performing a control task (e.g. maintaining a certain altitude) within a motion simulator showed impaired ability to perceive linear and angular movements when exposed to the additional mental workload of an auditory task. Cognitive factors in everyday life are, of course, not constant but vary depending on concurrent activities, prior knowledge, etc. In this thesis, the effect of cognitive factors on self-motion perception is addressed by measuring the human ability to detect motion while engaged in a driving task in a motion simulator (see section 1.4.3 and chapter 6).

1.3 THE CYBERNETICS APPROACH TO SELF-MOTION PERCEPTION

Cybernetics is a multidisciplinary approach to the study of self-regulatory systems - namely those that are able to generate changes in their environment which, in turn, are fed back to the system, altering its status and influencing future actions. By applying techniques from fields such as control theory, engineering, evolutionary biology, neuroscience and psychology, cyberneticists investigate the structures and capabilities of systems that include physical, biological and cognitive systems. When applied to perception, the cybernetics approach considers the brain of biological organisms as a complex control system where subcomponents can be isolated and individually investigated. Such an approach includes using system theory and psychophysics to model how the brain infers the physical world from sensory signals and generates actions to successfully interact with the world.

This thesis fits in the branch of neuroscience that applies the cybernetics approach to the study of self-motion perception. Specifically, this thesis focuses on psychophysical experiments that *objectively* quantify the human ability to perceive self-motion changes, or in other words the human sensitivity to self-motion. Objective sensitivity measures such as perceptual thresholds, discussed in the next session, are a key component for the development of cybernetics models that aim at quantifying perception and action in response to self-motion.

1.3.1 PERCEPTUAL THRESHOLDS: QUANTITATIVE MEASURES OF SELF-MOTION PERCEPTION

Sensory systems encode physical stimuli by using action potentials (i.e. spikes) traveling along the neurons from the sensory afferents toward the higher processing centres of the CNS (Kandel et al. 2000). Because the occurrence of an action potential is a probabilistic event, the number of spikes per second associated to a specific property of a physical stimulus is not constant but rather varies over time and over different repetitions (Faisal et al. 2008). The probabilistic nature of perception is grounded in this neural variability. This explains why different stimuli are sometimes indistinguishable for a human observer, or why repetitions of the same stimulus are sometimes perceived as different from each other. To understand how the central processing of sensory information leads to the subjective representation of physical stimuli, neuroscientists often rely on a class of theoretical and experimental methods called psychophysics (Kingdom and Prins 2010).

1.3.2 EXPERIMENTALLY MEASURING PERCEPTUAL THRESHOLDS

Widely exploited in the study of human self-motion perception are those psychophysical methods based on signal detection theory. Signal detection theory, also known as detection theory or decision theory, is the application of statistical tests to the problem of detecting a signal in the presence of noise (Green and Swets 1966; Kingdom and Prins 2010). By systematically studying the human ability to perform signal detection tasks it is possible to derive objective measures such as perceptual thresholds, which constitute a quantitative contribution to the study of the neural processes underlying self-motion perception. Experimental studies estimating perceptual thresholds can be divided into two main categories: estimation of absolute thresholds (the smallest detectable *level* of a stimulus) and estimation of differential thresholds (the smallest detectable *change* in stimulus intensity). A typical trial of a psychophysical self-motion study is structured as follow:

1. A motion stimulus, designed either prior to the experiment or generated online, is presented to the participant. Among the features of this motion stimulus (e.g. its direction, frequency, intensity, etc.), at least one of them is relevant for the specific question that the experiment is designed to address and might be manipulated (i.e. varied) across different trials. In psychophysics, the

manipulated variable is often referred to as the *independent variable* (Gescheider 1997; Kingdom and Prins 2010).

2. After stimulus presentation, the participant is required to make a decision that, ideally, is only based on the motion sensation evoked by the physical motion. For instance, the participant could be asked to report whether the motion was to the left or to the right or to align a pointer towards the origin of the motion.

Because perception is a probabilistic process, several redundant repetitions of the same stimulus are necessary to associate the probability of a participant's response (i.e. the *dependent variable*) to the physical manipulation. Moreover, different levels of the manipulated property of the physical stimulus need to be tested in order to investigate how the manipulation affects response probability. A perceptual threshold can eventually be defined as the level of physical manipulation of the stimulus required by the participant to respond with a given level of accuracy (e.g. 70%). A typical example of a psychophysical experiment and its outcome is provided in Figure 2.

In psychophysical experiments the stimulus designed by the experimenter is only one of several factors (or variables) that could affect the participant's responses. For instance, participants' responses also depend on prior knowledge, expectations and/or response strategies (sometimes referred to as *intervening variables*). This eventually results in noise in the experimental data. It is therefore of crucial importance to develop and employ experimental paradigms for stimulus presentation and data collection that minimize the level of noise in the data. Two among the most common experimental paradigms in psychophysics, the two-interval two-alternative forced choice (2IFC) task and the one-interval two-alternative forced choice (2AFC) task, were repeatedly employed throughout this thesis work and are introduced and discussed in the next paragraphs.

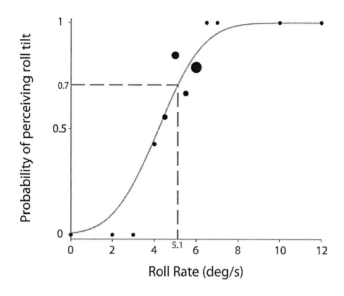

Figure 2 *Typical example of the results of a psychophysical experiment where a participant had to detect a physical tilt to the side (roll motion) in the experienced motion profile. The roll rate intensity of the stimulus (independent variable, x axis) was manipulated after every trial. The probability of detecting the roll (dependent variable, y axis) is a function of the roll rate intensity of the stimulus and is calculated for the tested roll rate intensities as the ratio between the number of correct detections over the number of times a particular stimulus intensity was presented (black dots). In this example a parametric function is fitted to the data and a stimulus intensity of 5.1 deg/s, corresponding to an estimated probability of 0.7 of correct signal detection, is chosen as the perceptual threshold. Figure adapted from Nesti et al. (2012).*

In a 2AFC task, participants are provided at every trial with one stimulus presentation that they have to respond to. For instance, they could be asked whether they perceived the motion or not (yes/no task or detection task) or, in presence of bidirectional stimuli, they could be asked to report the motion direction (e.g. left/right task or direction discrimination task). While, in a direction discrimination task, there is no reason to believe that participants report one direction more often than the other, detection tasks are inherently asymmetric since participants might be more concerned about not detecting a signal rather than reporting a signal when no signal was present (Merfeld 2011). This is usually referred to as a perceptual bias and is caused by the internal criterion that participants rely on to make their decision (Kaernbach 1990). Perceptual biases are an issue that deserves serious consideration when performing 2AFC tasks.

In a 2IFC task, participants are provided at every trial with two stimulus presentations and are then requested to perform a *relative* comparison (e.g. report the stronger of the two motions). This overcomes the issue of not being able to control for the internal criterion. However, choosing 2IFC over 2AFC has certain drawbacks. First, presenting 2 stimuli at every trial results either in a drastic increase of experiment duration (approximately double) or in a drastic decrease of collectable data, potentially leading to unfeasible or inconclusive experiments. Moreover, it could be challenging for the participants to compare two stimuli that are distant in time, either due to long pauses between the 2 stimulus presentations or because of long lasting stimuli. It is reasonable to expect that overall longer trials increase the chances of observing undesired effects in the data, such as those from presentation order or observer's fatigue.

Depending on the circumstances, both paradigms were employed in the experiments of this thesis work. A 2IFC was generally preferred because it is immune to any variability caused by participants' internal criterion. When the length of the stimuli prohibited 2IFC paradigms, a variation of the 2AFC paradigm (Kaernbach 1990) specifically designed to overcome issues related to the internal criterion was used.

1.4 THESIS OVERVIEW

1.4.1 AIMS

Systematic studies of human self-motion perception began in the 19th century (Mach 1875) and has received, since then, continuously increasing interest. In part, this is certainly due to the rise of human-piloted ground and aerial vehicles whose control is tightly related to self-motion perception. A wide range of biomechanical, neurophysiological and psychophysical studies have focused on important aspects of self-motion perception, such as sensory transduction of physical stimuli, propagation of neural signals within the CNS, multisensory integration and behavioural responses. The continuous improvements in the experimental hardware have undoubtedly benefitted this field immensely: compared to the technology available at the beginning of the 20th century, motion simulators have significantly extended their range and precision of reproducible inertial stimuli. Moreover, current virtual reality technologies allow for immersive visual environments and realistic physical models of the world and of the simulated vehicles. Overall, the study of self-motion is an active and dynamic research field where fundamental studies are conducted and merged into more general theories and models. Ultimately, self-motion perception research is expected to provide accurate descriptions of the perception of the complex motion patterns experienced in everyday life.

In line with the overall research directions of the field, the goal of this thesis is to investigate human self-motion perception in a realistic context, with the aim of better understanding the influence of sensory and cognitive factors on self-motion perception. This is done through a series of experiments conducted in the CyberMotion Simulator, a modern and highly performing motion simulator that is capable of providing motion stimuli that closely resemble the intensity, complexity and richness of several everyday scenarios. First, the question of how motion intensity affects human motion sensitivity is addressed. The ability of human participants to discriminate between two movements of different intensity is investigated for different motion ranges of translations and rotations in presence of inertial only or visual only cues. Analytical functions (defined in this work as perceptual laws) are proposed to relate the intensity of the motion stimulus with the smallest perceivable change in stimulus intensity and are identified based on experimentally collected data. Next, perceptual laws are identified in conditions where both visual and inertial information are available, as it is the case for the great majority of self-motion stimuli experienced in everyday life. The systematic comparison of human sensitivity to unimodal (i.e. visual-only or inertial-only) and multimodal (i.e. visual-inertial) motion cues further allows validation of multisensory integration theories as well as speculation on the neural mechanisms that underlie the discrimination of different self-motion intensities. In the last part of the thesis, experiments were developed and extended towards a more ecological scenario, by using a driving simulation as experimental framework. Here, motion stimuli include multimodal cues that ranged over different intensities but further increase in richness and complexity by combining linear and angular movements so to reproduce the experience of driving on a curving road. Moreover, the role of mental and physical workload of a driving task on self-motion perception is addressed.

In summary, this thesis begins with very controlled experiments where single contributing factors to self-motion perception are systematically studied and develops, one factor at a time, towards experimental conditions that closely resemble the complexity of the real world. The next two sections provide an overview of the experimental contents of this thesis, which are presented in greater details in the following chapters. An overall discussion is presented in section 1.5 to address the implication of the experimental results and indicate future research directions.

1.4.2 PART 1: SUPRA-THRESHOLD SELF-MOTION SENSITIVITY

In the context of sensory systems the term sensitivity is used with different meanings. Generally, the sensitivity of a sensory system (e.g. a weighing scale) is defined as the minimum intensity of an input signal that is required to produce an output signal that satisfies a specified criterion. For a sensory system whose output is discrete, such criterion could be a change of level on the output scale (Figure 3, left). Sensitivity would then be defined as the minimum increase in the sensory input as is required to elicit a change of the discrete output. For example, the sensitivity of a digital weighing scale is the minimum weight that needs to be added on the scale in order to read a value change on the display. Another common criterion, typically employed in sensory systems affected by noise (Figure 3, right), is based on the relative amount of signal compared to noise (i.e. its signal-to-noise ratio). Sensitivity would then be defined as the minimum intensity of input signal that generates an output with the specified improvement of the signal-to-noise ratio. Note that, strictly speaking, from this definition of sensitivity it follows that low sensitivities reflect small input changes. However, it is common practice to refer to systems which are able to detect relatively small input changes as being highly sensitive, i.e. sensitivity is commonly considered inversely proportional to the smallest detectable input changes. Although this definition of sensitivity is widely employed, it is not uncommon to see the term "sensitivity" (or alternatively responsivity) when referring to the ratio between changes in the output and changes in the input of the sensor, i.e. the derivative of the output with respect to the input. Furthermore, sensitivity, as defined at the beginning of this paragraph, is sometimes also referred to as the resolution of the sensor. Throughout this thesis, the term "sensitivity" is consistently used in a more abstract sense and refers to the ability of a human observer to discriminate between different motion profiles. Quantification of such ability is provided by experimentally measuring differential thresholds (introduced in section 1.3), i.e. the smallest change in stimulus intensity that can be detected by a human observer in a given percentage of observations (Gescheider 1997). This terminology is consistent with the psychophysical literature and should prevent confusion when results of this thesis work are compared with previous studies. Furthermore, throughout this thesis the expression "high sensitivity" is associated to relatively small

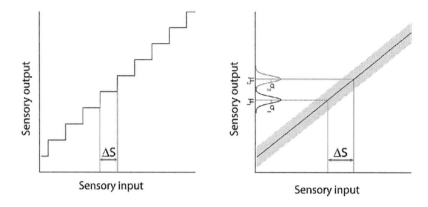

Figure 3 Left: *The input-output curve of a discrete sensory system. Responses of this system to a changing physical quantity are separated by finite intervals along the "sensory output" dimension. The smallest change ΔS of the measured physical quantity that elicits a change in the discrete output of the sensor is called, depending on the context, sensitivity, resolution or differential threshold.* **Right:** *In a system affected by noise, the value read from the sensor in response to a stimulus presentation is drawn from a distribution of possible outputs. Therefore, if two stimuli are close together it can happen that the sensor reports a higher output for the weaker stimulus. The sensitivity (or resolution, or differential threshold) of such system can be defined as the smallest change in stimulus intensity ΔS for which the sensory response satisfies $\mu_2 / \sigma_2 = k * \mu_1 / \sigma_1$, i.e. the signal-to-noise ratio in the sensor response improves by a factor k. For every k, it possible to compute the probability of correctly detecting/discriminating a change ΔS in stimulus intensity.*

differential thresholds, whereas low motion sensitivity refers to impaired ability to discriminate different motion intensities (i.e. large differential thresholds).

Using psychophysical techniques it is often possible to obtain quantitative measures for the sensitivity of human perception. As discussed in section 1.3, two widely employed perceptual measures of sensitivity are absolute thresholds (i.e. the smallest detectable level of stimulus) and differential thresholds (i.e. the smallest detectable change in stimulus intensity). Identifying the relationship between sensitivity and stimulus intensity requires differential thresholds for a set of stimuli covering a range of different intensities to be measured. This type of studies was initiated by Weber in the 19th century (Fechner 1860), who measured differential thresholds for the perception of objects' weight. His seminal works gave rise to a wave of studies over the last century and differential thresholds are, nowadays, still actively investigated in a variety of research fields including self-motion perception (Zaichik et al. 1999; Mallery et al. 2010; Naseri and Grant 2012). Findings generalize remarkably well across several different perceptual modalities (cf. Teghtsoonian (1971) and Table 1) and reveal that differential thresholds are not independent from stimulus intensity but rather increase for increasing stimulus intensities. Weber proposed a linear law (later named by Fechner "Weber's law" (Fechner 1860)) as a model to relate stimulus intensities and differential thresholds.

More recent work (cf. Teghtsoonian (1971)) proved that Weber's law described perception well only for stimuli of moderate intensity. Indeed, when considering ranges that include the lowest and highest stimulus intensities perceivable by humans during everyday life, deviations from Weber's law can be observed (Teghtsoonian 1971) and a power law captures such changes in DTs better (Guilford 1932). As extensively discussed in the proceeding of this thesis, sensory systems showing stimulus-dependent sensitivity (like the ones described by Weber's Law) are by definition nonlinear systems, meaning that their output is not linearly proportional to their input. In contrast, a linear system maintains a

Continuum		k
Light Intensity		0.079
Sound Intensity		0.048
Finger Span		0.022
Lifted Weight		0.020
Line Length		0.029
Taste (salt)		0.083
Electric Shock		0.013
Vibration (fingertip)	60 Hz	0.036
	125 Hz	0.046
	250 Hz	0.046

Table 1 *Weber's perceptual law has been identified for several different perceptual continua by experimentally measuring differential thresholds and fitting the Weber's constants (k) to the linear model ΔS = k * S, where ΔS is the differential threshold, S the stimulus intensity and k the Weber's constant. Adapted from Teghtsoonian (1971).*

constant level of sensitivity over the entire input range. The two input-output curves in Figure 3 are both representative of a general linear system.

Systematic studies on self-motion sensitivity did not appear until recent times, partly because of the technical challenges of using motion simulators to accurately reproduce motion stimuli of high intensity. In 1999 Zaichik and colleagues used a motion platform to measure human differential thresholds for translations in the dark in the range of $0 - 0.6$ m/s^2 (Zaichik et al. 1999) and found them to be consistent with Weber's law. These intensities are very common in daily activities such as walking, but the human body is also exposed to higher intensities, particularly from running or traveling in motor vehicles. Wider ranges of head-centred rotations (0-150 deg/s) and longitudinal accelerations (0-2m/s^2) were also investigated by Mallery et al. (2010) and by Naseri and Grant (2011) respectively, and the obtained perceptual laws again supports the nonlinearity of self-motion perception. This thesis contributes to this line of research by identifying for the first time perceptual laws for the perception of heave translations in darkness, yaw circular vection and yaw rotations in presence of congruent visual and inertial information (i.e. visual and inertial cues of equal intensity and opposite direction). Furthermore, the study by Mallery et al. (2010) on yaw rotation sensitivity in darkness is replicated. A considered design of the studies on yaw circular vection, yaw sensitivity in darkness and yaw sensitivity to visual-inertial yaw rotations further allows investigation of the multisensory integration mechanisms employed by the CNS in presence of congruent visual-inertial motion cues. This thesis also touches a fundamental technical issue concerning the use of motion simulators to present carefully designed motion stimuli. Indeed, important distortions in the motion stimulus are often unavoidable due to the mass and complexity of the experimental setup and a thorough analysis of the reproduced stimuli is therefore essential to prevent possible erroneous interpretations of the experimental results.

These studies are individually summarized in the next paragraphs. For each study, greater details can be found in the thesis chapter reported in square brackets at the beginning of each paragraph.

The importance of stimulus noise analysis for self-motion studies [chapter 2]

Investigating the neural and cognitive mechanisms of self-motion perception often requires motion stimuli presented by means of a motion simulator. Despite repeatedly raised concerns by neuroscientists about simulator-introduced noise, there was no study yet that thoroughly examined and discussed the implications of such issue. In the work presented in chapter 2, easily grasped graphical and statistical techniques are developed for the analysis of motion stimuli commonly employed in self-motion studies. The analyses and their relevance are illustrated using motion stimuli from the study on heave sensitivity described in chapter 3. Inertial recordings from the simulator are analysed using quantitative measures of noise, commonly employed in the design and validation of motion simulators (AGARD 1979), and standard statistical hypothesis tests. Results show direction-dependent noise and nonlinearity related to the motion intensity over a range of 0-2 m/s^2, with the Signal-to-Noise Ratio improving for increasing motion intensities. This dependency needs to be seriously considered as it implies that manipulating stimulus intensity affects both the intensity and the quality of the reproduced motion and might generate confounds. Overall, the proposed analyses provide a valuable set of tools for the interpretation of neurophysiological and behavioural responses, allowing for dissociation between simulator and physiological noise as well as for meaningful comparisons across the existing literature. The quantitative approach proposed in this chapter for quantifying the noise in the experimental motion stimuli is expected, due to its generalizability and comprehensibility, to find immediate and enduring application in a broad variety of neurophysiological, physiological and psychophysical studies on animals and humans, including neural processing of sensory information, eye movements, balance disorders and dynamic vehicle simulation.

Human sensitivity to vertical self-motion [chapter 3]

Among the unexplored degrees of freedom for which sensitivity can be measured, heave is of particular interest due to the constant presence of gravity that is indistinguishable from an over-imposed inertial acceleration. Despite the fact that gravity is not perceived in an observer at rest as an upward motion, the presence of a non-zero pedestal stimulus could influence self-motion sensitivity and lead to perceptual asymmetries. In chapter 3, human differential thresholds for both upward and downward translations were measured over a broad range of motion intensities (0-2 m/s^2) using single-cycle sinusoidal acceleration profiles. In both directions differential thresholds are found to increase with stimulus intensity. As discussed previously, this constitutes a nonlinearity in the perception of self-motion because sensitivity does not remain constant throughout the investigated motion range. Comparison of different models based on information criteria reveals that a concave power law, better than a Weber's law, describes the increase of differential thresholds with increasing stimulus intensity. This deviation from Weber's law might reflect the necessity of humans to precisely estimate motion even at high motion intensities, in order to maintain balance and prevent falls. Lower differential thresholds are measured for downward as opposed to upward motion, an asymmetry not reported in absolute threshold studies or in neurophysiological recordings from the vestibular afferent fibres. Analysis of the stimulus noise (see chapter 2) revealed that downward stimuli contain less noise than upward stimuli, a stimulus asymmetry that might explain the direction asymmetry of measured differential thresholds. An alternative explanation for higher downward sensitivity could be again related to adaptations to avoid falling.

Self-motion sensitivity to visual yaw rotations in humans [chapter 4]

Despite the well-established role of visual cues in self-motion perception (von Helmholtz 1925; Gibson 1950), little effort has been dedicated to measuring differential thresholds for visual self-motion cues. One of the main challenges of such study is certainly related to the intrinsic ambiguity of visual cues which might stem from both object motion and self-motion. The work described in chapter 4 includes the development of a novel methodology for the study of vection sensitivity that carefully prevents this possible confound and ensures that all the visual motion is perceived as self-motion. Differential thresholds for vection as evoked by constant visual rotations around the vertical (yaw) axis were measured over a range of different rotation intensities (0-60 deg/s) and found to significantly increase with the velocity of visual motion. This nonlinearity in the perception of self-motion from visual cues closely resembles findings from self-motion studies in darkness conducted within this thesis and elsewhere (Zaichik et al. 1999; Mallery et al. 2010; Naseri and Grant 2012). Furthermore, it was observed in this study that the time necessary for full vection to occur (i.e. the time between stimulus onset and when participants experienced self-rotation in a stationary environment) is independent from stimulus velocity and shorter after previous exposure.

Human sensitivity to inertial yaw rotations [chapter 5]

Human sensitivity to head-centred yaw rotations in darkness was previously investigated by Mallery et al. (2010), who also reported that differential thresholds increase in accordance with a power law type of perceptual law. A replication of these results is included in the experiments of this thesis as it serves different purposes. First, measuring both sensitivity to inertial-only and visual-only motion cues (chapter 4) using the same experimental procedure, setup and participants allows for comparison of sensitivity of the two main types of sensory information available to the CNS for estimating self-motion. Furthermore, a third experiment, again employing the same participants, setup and design, was conducted to measure differential thresholds in presence of congruent visual and inertial cues and investigate therefore the process of multisensory integration for self-motion perception (more details are provided in the next paragraph and in chapter 5). An additional argument for replicating the study of Mallery and colleagues is to address the question of how well results might generalize over different participants and different simulators. Differential thresholds were, therefore, measured for sinusoidal yaw rotations in darkness at different peak intensities ranging between 15 and 60 deg/s. The analytical perceptual law that emerges from the data collected in this thesis resembled the one from Mallery and colleagues remarkably well. Indeed, the exponent of the power law, which reflects the rate at which thresholds increase with stimulus intensity, is almost identical (0.36 compared to 0.37). The overall higher thresholds found in the present work, reflected in the higher gain, are likely due to differences in the simulators. Measured differential thresholds for yaw rotations in presence of inertial-only and visual-only (chapter 4) motion cues are not significantly different, suggesting a common neural mechanism acting on the internal representation of self-motion within the CNS.

Human sensitivity to head-centred visual-inertial yaw rotations [chapter 5]

Whereas selectively presenting visual-only or inertial-only motion cues is essential for disentangling the contribution of different sensory systems to self-motion perception, multimodal stimuli claim higher ecological validity. In the vast majority of everyday scenarios the CNS can rely on multiple sources of information, primarily from the visual system and the inertial sensory systems (vestibular and somatosensory). It is widely acknowledged that, when estimating physical stimuli such as self-motion, humans can achieve higher sensitivity when multiple sources of sensory information are combined in contrast to when sensory cues are presented individually (Ernst and Banks 2002; Doya et al. 2007). The work described in chapter 5 aims at providing a quantitative description of self-motion sensitivity in the presence of combined visual and inertial cues. This is because little is known about visual-inertial perceptual integration and the resulting self-motion sensitivity over a wide range of motion intensity. Differential thresholds for yaw rotations in the range of 15-60 deg/s were measured using the same participants, procedure and stimuli employed in the study on human sensitivity to head-centred yaw rotations in darkness (see previous paragraph and chapter 5). In this experiment, visual information congruent with the inertial stimulus was provided. Results indicate that differential thresholds increase with stimulus intensity following a trend described well by a power law. However, differential thresholds for yaw rotations measured here under visual-inertial cues are not significantly lower than those measured for visual-only and inertial-only motion cues (see chapter 4 and 5). This suggests that combining visual and inertial motion cues in a yaw intensity discrimination task does not lead to improved self-motion sensitivity, as predicted by general theories on multisensory perception.

1.4.3 PART 2: PERCEPTION BASED MOTION SIMULATION

In line with the goals of this thesis, part 2 constitutes an active effort towards gaining better understanding on how self-motion from everyday life is perceived. As compared to part 1, the experiments in part 2 are characterized by an overall more "ecological" design. In an ecological scenario motion stimuli closely resemble the variety, richness and complexity of the real world, providing visual-inertial cues that involve combinations of motions in different degrees of freedom as well as increased mental and physical workload. A driving scenario contains several of these elements and represents an ideal framework for the investigation of how sensory and cognitive factors can contribute to self-motion perception in ecological situations.

Besides its high ecological content, the study of self-motion perception during driving is of great interest for the field of vehicle motion simulation, where such knowledge is exploited in order to best recreate within a motion simulator the motion experienced on the real vehicle (i.e. the so-called motion cueing). The large spectrum of challenges in motion simulation covers aspects such as the quality of the visual scene, the ergonomics of the environment, the need of an accurate vehicle model and the limited motion range (Fisher et al. 2011). Among these challenges, the latter can be addressed by using knowledge of human self-motion perception in an attempt to reproduce the perceived rather than the physical motion of the vehicle. According to sections 1.2 and 1.3 it should be clear that perception is not perfect but is subject to ambiguities from different sources such as sensor dynamics, multisensory integration and physiological noise. In motion simulation, a common "trick" that exploits a perceptual ambiguity is the so-called tilt-coordination technique, present in the vast majority of algorithms for motion cueing. This technique effectively simulates low frequency accelerations by relying on the tilt-translation ambiguity (described in section 1.2.2 and illustrated in Figure 4), which is caused on a perceptual level from an insensitivity in discriminating between the resulting forces from body tilt angle and linear acceleration. Indeed, both inertial accelerations when sitting upright and body tilt with respect to gravity causes in the driver a sensation of being forced into (or away from) the driver's seat. This illusion occurs because different combinations of linear accelerations and static body tilt result in similar gravito-inertial forces on the inertial sensory systems.

To solve the tilt-translation ambiguity, the CNS makes use of visual orientation cues (Groen and Bles 2004) and inertial rotation cues (Angelaki and Yakusheva 2009). A widely employed motion cueing algorithm is the washout algorithm (schematically illustrated in Figure 5). Such algorithm includes low-pass filtering of the linear accelerations of the vehicle to isolate those sustained motion components that likely exceeds the limits of the simulator workspace. Driver's perception of these sustained

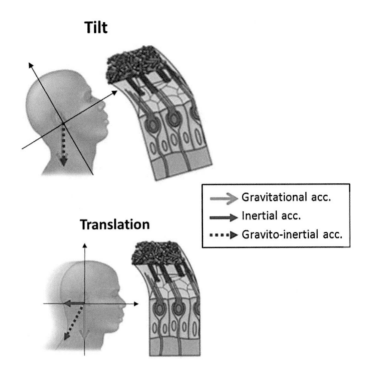

Figure 4 *Acceleration sensors such as the otolithic organs respond to the vectorial sum of gravity and inertial accelerations (i.e. the gravito-inertial vector) and can therefore lead to ambiguity between a static head tilt (top figure) and a linear acceleration with the head upright (bottom figure). Indeed, the same orientation of the gravito-inertial vector with respect to the head resulting from a linear acceleration is achievable in a static head tilt position. Note that, although the intensity of the gravito-inertial vector is always slightly higher in presence of linear acceleration compared to static head tilt, this is usually not sufficient to solve the ambiguity.*

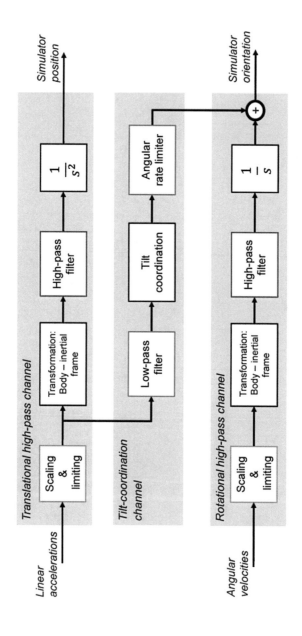

32

Figure 5 *(preceding page) Schematic representation of a classic washout filter for motion cueing (Nahon and Reid 1990). The accelerations and angular velocities experienced by the driver in the real (or simulated) vehicle are first scaled and saturated (orange blocks) to account for the physical limits of the simulator. A change of reference frame (black blocks) is also required as most (if not all) motion simulators are controlled using inertial coordinates (i.e. relative to the world rather than relative to the driver's head). Next, both linear and angular motions are high-pass filtered (red blocks) to avoid low-frequency movements (i.e. movements sustained for long times) which easily run into the limits of the simulator workspace. Tilt-coordination acts in rescue of the discarded low-frequency translations and computes the angular commands necessary to exploit gravity in the simulation of sustained linear accelerations. This perceptual trick requires saturation of the tilt rate below human perceptual threshold (green block) to prevent the false cues introduced from being perceived, spoiling the motion illusion. An additional feature of washout filters consists in continuously repositioning the simulator towards the centre of its operational workspace, where the dexterity is higher. This is achieved by wisely designing the high-pass filters (red blocks) so that the signal controlling the simulator position and orientation is "washed out" (hence the name of the algorithm), possibly using sub-threshold movements. The blocks "1/s" and "1/s^2" at the end of the translational and rotational channels represent respectively the operations of integration and double integration written in terms of the Laplace transform variable "s". These operations are required to transform the filtered signals in position and orientation commands for the motion simulator.*

components is, however, preserved by tilting the simulator cabin while ensuring that the tilt-translation ambiguity is resolved in favour of the feeling of linear acceleration. This is achieved by providing visual cues congruent with linear self-motion and by tilting the simulator cabin below human perceptual threshold.

It is not trivial to make a sensible choice about the tilt constraints that are required to maintain an unperceivable tilt during tilt-coordination. Indeed, as already extensively discussed in this chapter, human sensitivity to angular rotations is affected by a variety of factors such as intensity, frequency and direction of the motion as well as visual information and cognitive factors. A conservative choice would be to rely on the lowest thresholds measured in literature, i.e. the smallest perceivable rotation intensity. For instance, Zaichik et al. (1999) measured thresholds as low as 0.5 deg/s for participants who were fully concentrated on detecting the direction of a pure roll (leftward/rightward tilt) and of a pure pitch (forward/backward tilt) motion. Constraining tilt-coordination to never exceed such a tilt rate level would surely prevent participants from perceiving the tilt, as the additional factors present in a driving simulation (e.g. co-occurring translations, active driving task) only have deteriorating effects on tilt sensitivity. On the other hand, the downside of choosing such tilt rate limitation is a serious impairment of tilt-coordination efficiency: reaching the desired tilt angle (i.e. aligning gravity with the desired gravito-inertial vector) takes too long and the correct acceleration feeling is achieved only with a long delay. However, experimental evidence suggests that the human tilt threshold might increase under certain conditions, for instance in the presence of a visual stimulus that does not tilt (Groen and Bles 2004) or when participants are engaged in additional tasks (Hosman and van der Vaart 1978) or again when rotational and translational motion are combined (Zaichik et al. 1999). These and other factors are present in a driving task, suggesting that higher tilt rate saturation values might simulate low frequency accelerations better without spoiling the simulation realism.

The work conducted in part 2 of the present thesis aims to measure how individual sensory and cognitive factors that are typical of ecological scenarios might affect human perception of self-motion. Measuring tilt rate perceptual thresholds in the context of a car driving simulation has the twofold benefit of quantitatively answering this question while, at the same time, providing directly applicable perceptual knowledge to the field of motion cueing. In chapter 6, roll rate detection thresholds were measured under different combinations of roll motion, lateral motion, visual information and the presence of an active driving task. Results show that sensitivity is highest for pure roll stimuli in darkness. Thresholds increase approximately 6 times when the lateral acceleration of the car is also reproduced, a result that qualitatively agrees with previous findings (Zaichik et al. 1999). Providing visual information, incongruent with the physical roll, did not affect sensitivity but the measured threshold again qualitatively agrees with studies from the literature (Groen and Bles 2004). Interestingly, the attentional load requested by an active driving task only affected a group of drivers whereas the rest maintained similar sensitivity. Experimental data suggests that, when the driving simulation is perceived as realistic, drivers can better concentrate on the motion and maintain higher sensitivity as a result.

1.5 DISCUSSION AND FUTURE WORK

Self-motion perception is a nonlinear process that involves the contribution of sensory and cognitive factors for estimating self-motion through the environment. In this thesis, human self-motion perception was investigated focusing on aspects of self-motion that are typical in everyday scenarios yet have received relatively little attention in the literature. First, the question "how sensitive are humans to changes in self-motion intensity?" was addressed at different intensities of visual and inertial motion cues, presented independently or congruently combined. Sensitivity is a fundamental aspect in the engineering of any system that is required to process sensory information over wide ranges of stimulus intensity. It directly affects the actions that such system can or cannot perform. Although daily activities expose humans to wide ranges of self-motion intensities (for example yaw rotations during locomotion range between 0 and 400 deg/s (Grossman et al. 1988)), research efforts on self-motion perception have mainly focused on other important aspects such as frequency dependencies, direction dependencies and asymmetries, aging and vestibular disorders. Next, the effect of a lateral acceleration on the sensitivity to roll motion was investigated. Combination of different lateral and angular movements are clearly a much more natural stimulus compared to those movements in a single degree of freedom commonly employed in basic psychophysical experiments. The current results constitute a step forward towards the understanding of self-motion perception in everyday life situations. The effect of incongruent sensory information on tilt sensitivity was also investigated. Although natural self-motion provides congruent visual and inertial information in the vast majority of the cases, knowing the effect of an incongruent visual stimulus on the perception of lateral tilt is desirable for the design of any virtual environment. In virtual environments, visual cues originate from a display rather than the real world and, with appropriate knowledge of self-motion perception, can be adapted at convenience to manipulate the self-motion perception evoked in the observer, with great applied benefits, such as in the field of motion simulation. Eventually, an active driving simulation was designed to study the effect of an active task on tilt sensitivity. This experimental condition is very close to an everyday life situation where self-motion presents complex motion patterns with visual and inertial motion cues as well as the attentional demand of a common active task. This section briefly discusses the experimental results of this thesis (summarized in section 1.4) with respect to their implications for basic and applied research and outlines future directions for theoretical and experimental works.

The work conducted in this thesis provides convincing evidence that human sensitivity to self-motion, quantified by measuring differential thresholds over wide intensity ranges of linear and angular motion, is not constant. Instead, it decreases for increasing intensities of the physical stimulus. This behaviour is typical of nonlinear systems, since linear systems are characterized by constant sensitivity over the entire range of stimuli evoking a sensory response. Nonlinearities in perceptual processes seem to be a general property of the CNS, which extends to the perception of several classes of physical stimuli (Teghtsoonian 1971; Baird 1997). However, before concluding that differential thresholds measured in this thesis reflect nonlinearities in the perception of self-motion, possible confounds due to the mechanical noise of the simulator had to be addressed. Analysis of the linear and angular motion stimuli as reproduced by the CyberMotion Simulator revealed that their SNRs, an index of how faithfully the motion command is reproduced, increased with motion intensity. This proves that simulator noise cannot explain the decrease in participants' motion sensitivity, which is instead expected to increase for higher SNRs (Greig 1988). Increased uncertainty (i.e. higher differential threshold) in the discrimination of stronger compared to weaker stimuli must, therefore, reflect a source of noise that is physiological and internal to the CNS. Overall, the analyses performed on the noise of motion stimuli allow for straightforward dissociation between simulator and physiological noise and are applicable to the great majority of motion stimuli commonly employed in self-motion studies. This sets an important precedent for neurophysiological, psychophysical and behavioural studies that aim to inform the neural and cognitive processes of self-motion perception.

Important considerations emerge when results from self-motion perception studies are compared to studies on ocular reflexes and neurophysiology. The Vestibulo-Ocular Reflex (VOR) and the optokinetic reflex, responsible for countering head movements with eye movements and, thus, maintaining a stable view, do not show the nonlinear behaviour observed for perception. Instead, approximately constant level of precision and accuracy over different visual (Paige 1994) and inertial (Pulaski et al. 1981; Weber et al. 2008) motion intensities are reported, thus suggesting that the eye movements can be described as a linear process (Merfeld et al. 1993). Neurophysiological evidence for perceptual nonlinearities is not found in the primary afferent fibres of the vestibular organs (otoliths and semicircular canals) (Sadeghi et al. 2007; Massot et al. 2012; Jamali et al. 2013) nor in neurons of the vestibular nuclei (Dichgans et al. 1973; Henn et al. 1974; Waespe and Henn 1977), where average firing rates linearly depend on stimulus intensity (qualitatively resembling the right plot in Figure 3). It is, however, interesting to observe that in the vestibular nuclei (but not in the vestibular afferents) variability in the neuronal signal (i.e. number of action potentials per seconds) increases for stronger compared to weaker motion intensities (Dichgans et al. 1973; Henn et al. 1974; Allum et al. 1976; Waespe and Henn 1977; Massot et al. 2011). This increase in variability is expected to worsen discrimination performances (i.e. increase differential thresholds) even when the average sensory response maintain its linear characteristic. Future studies are still required to quantify the relationship between neural variability and stimulus intensity in a similar way as it was done in this thesis to identify perceptual laws. This can be done by computing "neuronal thresholds" (Angelaki et al. 2009), which represents the ability of an ideal observer to discriminate between different physical stimuli based on the activity of a single neuron. Perceptual differential thresholds and neuronal differential thresholds obtained from recordings in the vestibular nuclei and in higher processing centres of the CNS can then be compared, to localize the neuronal processes that are responsible for perceptual nonlinearities. Candidate areas are the parieto-insular vestibular cortex, the medial superior temporal area and the ventral intraparietal area (Yu et al. 2012).

Unlike in linear systems, the sensitivity of a nonlinear system can depend on the intensity of the physical stimulus. In the case of self-motion perception, higher sensitivity is reported for weaker compared to stronger motion intensities. Such nonlinearity has beneficial aspects and might, therefore, reflect a strategy of the CNS. Indeed the range of possible sensory responses (i.e. firing rate) is limited and imposes a trade-off between the system sensitivity and the range of stimulus intensities on which the system remains sensitive. If specific intensities are more important or more frequent than others, it is an efficient strategy to assign to these intensities higher sensitivity. In chapter 5 (see Figure 32) this idea is illustrated using the distribution of stimulus intensities recorded during a daily activity (i.e. jogging). Future work should further investigate the relationship between the statistics of stimuli intensities during every-day life and the corresponding human sensitivity. Moreover, evidence of perceptual learning in self-motion perception (Hartmann et al. 2013) suggests that the overall sensitivities as well as the shape of the perceptual laws might be altered by previously exposing participants to stimuli of different intensities. Such experiments might be used to validate computational models of *efficient coding* (Wei and Stocker 2013), which suggests a link between perceptual responses to a given stimulus and the underlying stimulus distributions. Moreover, perceptual nonlinearities might inspire a bionic approach to the engineering of artificial sensors. This would lead to better trade-off between sensitivity and sensor range in those cases where knowledge about the probability distribution of the physical quantity is known or can be envisaged.

The perceptual laws identified in the present thesis allow for a quantitative prediction of differential thresholds as a function of motion intensity, quantitatively informing about the uncertainty in the discrimination between motions of different intensity. Intuitively, it might seem that differential thresholds can also allow the construction of a "perceptual scale" relating physical to perceived intensity of the motion. For instance, a perceptual scale can be derived by assigning an arbitrary value P_i on the perceptual scale to a given stimulus intensity S_i. An increment of one perceptual unit (from P_i to P_{i+1}) is then assigned to the physical stimulus with intensity $S_{i+1} = S_i + d_i$, where d_i is the differential threshold experimentally measured for a reference stimulus S_i. This method of assigning to each differential threshold an increment of one perceptual unit is sometimes referred to as Discrimination scaling or Fechnerian integration (Gescheider 1997). Gustav Fechner was also the first psychophysicist who suggested that if differential thresholds obey Weber's Law ($d_i = k * S_i$) the underlying perceptual scale can be approximated by a logarithmic function (Fechner 1860). Although he further provided a mathematical derivation of the logarithmic law, this derivation has been subject to many controversies in the literature (cf. Luce and Edwards (1958)). There is, however, a serious problem in the method of Discrimination scaling due to the fact that differential thresholds alone do not allow to infer both the shape of the function relating physical to perceived stimuli (perceptual function) *and* the noise that is internal to the perceptual process (internal noise) (Kingdom and Prins 2010). This problem is illustrated by the examples in Figure 6: increasing differential thresholds (such as those observed throughout the chapters 3 to 5) might emerge from either a nonlinear perceptual function with constant noise (Figure 6, left) or a linear perceptual function where the noise increases with the intensity of the stimulus (Figure 6, right), which also results in an overall nonlinear response. Of course, the more general case where the perceptual function is nonlinear and the noise is not constant is also possible. Determining the perceptual function from the differential thresholds is, therefore, not possible unless information about the noise is also available. Important advances in the study of self-motion perception will derive from studies that are specifically designed to address these issues. An interesting scaling method already present in the literature that is theoretically able to measure both the shape of the perceptual function and its noise is the Maximum Likelihood Difference Scaling (Maloney and Yang 2003; Kingdom and Prins 2010). One possible benefit of applying such method to self-motion perception studies is the comparison of perceptual functions and the associated noise with neurophysiological studies, which would surely inform us about the origin of perceptual nonlinearities. For example, Massot et al. (2011) showed that there are neurons in the vestibular nuclei whose response to head-centred yaw rotations is on average linear and affected by stimulus dependent noise,

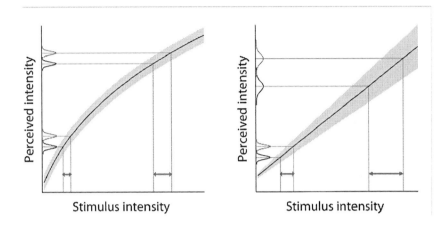

Figure 6 *Two different perceptual functions mapping the intensity of a physical stimulus to the perceived intensity. The perceptual function in the left diagram is nonlinear and less steep for higher compared to lower motion intensities. Consequently, changes in the lower compared to higher ranges of motion intensity are more readily detectable and bigger changes are therefore needed to obtain comparable discrimination performances (e.g. a given percentage of correct detection). The plot on the right also represents a system whose sensitivity is higher for weaker stimuli. However, the drop in sensitivity at higher stimulus intensities is here due to the increase of internal noise rather than to a nonlinearity of the perceptual function, which in this case is linear.*

qualitatively resembling the right panel in Figure 6. If scaling methods such as the Maximum Likelihood Difference Scaling or the Partition Scaling (Kingdom and Prins 2010) were to find a nonlinear perceptual function, it would be reasonable to expect that the neural processes responsible for this nonlinearity are located deeper along the neuronal paths involved in self-motion perception.

Models of visual-inertial sensory integration have been developed in the past to describe eye movements (Robinson 1981), balance responses (van der Kooij et al. 2001; Kuo 2005), motion sickness (Oman 1982), body orientation perception (Borah et al. 1988) and self-motion perception (Bos and Bles 2002; Zupan et al. 2002; Newman 2012). These models, validated through many psychophysical experiments, rely on the assumption that perception is linear over the range of possible motion intensities once absolute threshold is overcome. In other words, it is assumed that our sensitivity to supra-threshold self-motion is not affected by motion intensity. Results of this thesis disprove this assumption and show that sensitivity decreases at higher intensities. Moreover, the current results also show that motion complexity (i.e. combining motion in different directions) and cognitive factors have an effect on self-motion perception. Accounting for these factors would certainly improve models accuracy in situations with complex visual-inertial stimuli and attention demand, which are typical of everyday life. Future modelling work should pursue several aims. First, perceptual nonlinearities that emerged from the studies on differential thresholds need to be implemented in a computational model to account for the nonlinear relationship between human sensitivity and stimulus intensity. Results presented in chapter 5 showed similar differential thresholds for visual-only, inertial-only and visual-inertial motion cues, suggesting that perceptual nonlinearities could be implemented in a self-motion perception model as a "linear-nonlinear cascade" (Figure 7) after the interaction of visual and inertial sensory information. As discussed in the previous paragraph, the implementation of perceptual nonlinearities could be achieved by assuming constant internal noise (i.e. noise internal to the perception process) and fitting a logarithmic function, as suggested by Fechner (1860), or a power function, as suggested by Stevens (1957), to the perceptual scale derived from differential thresholds. Alternatively, experiments could be designed using the method of Maximum Likelihood Difference Scaling to systematically address the estimation of both the shape of the perceptual function and the internal noise. A secondresearch direction should focus on the identification of analytical laws that describe how motion sensitivity in one degree of freedom is affected by simultaneously occurring motion in the other degrees of freedom (cross-modal perceptual laws). Evidence that rotational sensitivity drops when translational and rotational motions are combined is reported in chapter 6 of this thesis and by previous researchers (Zaichik et al. 1999), although the cause for this sensitivity loss remains unclear and could possibly be due to a distraction effect (similar to the effect that a loud sound would have). Nevertheless, real-life scenarios often provide combined translations and rotations and cross-modal perceptual laws are, therefore, of great importance for modelling self-motion perception. Lastly, in this thesis the effect of an active task on self-motion perception is addressed in the context of a driving simulation. In chapter 6 it is speculated that the level of immersion in the driving simulation determines the sensitivity of the drivers. Although an increase in cognitive workload can be expected when moving from passive to active driving simulation, collected questionnaires did not provide evidence for this. Future work should focus on using physiological measures such as heart rate variability, skin conductance and ElectroEncephaloGraphy (EEG) in real-life scenarios. Based on previous literature, reviewed in detail by Cain (2007), these physiological parameters could correlate well with the amount of workload and attention demanded by active tasks and could, therefore, be included, due to their quantitative nature, in computational models of self-motion perception.

Figure 7 *(preceding page) A linear-nonlinear cascade model for the perception of angular self-motion. The two blocks on the left, V(s) and I(s), respectively represent the dynamics of the visual and inertial sensory systems using a notation typical for transfer functions, i.e. mathematical representations commonly employed in self-motion perception models to describe the input-output relationship of sensory systems[3]. The block on the right converts the integrated visual and inertial signal to perceived angular velocity and accounts for the nonlinear relationship between physical and perceived motion intensity. Concerning the mathematical integration of visual and inertial sensory signals (the operator ⊕), different solutions are possible. For instance, visual and inertial signals could integrate according to the theory of maximum likelihood estimation (Doya et al. 2007), although in chapter 5 of the present thesis predictions based on such theory were not confirmed by differential thresholds measured experimentally. A similar perception model was previously suggested by Zaichik et al. (1999) using a linear-nonlinear cascade where the visual dynamics were absent and the perceptual nonlinearity (a discontinuous power function) directly followed the dynamics of the inertial sensory systems. In a related context, a linear-nonlinear cascade model was suggested to account for the attenuated sensitivity of neurons in the vestibular nuclei to low frequency stimuli when low and high frequency stimuli are presented concurrently (Massot et al. 2012).*

---

[3] A short introduction to transfer functions can be found in Soyka et al. (2011), a complete coverage of the topic can be found in basic textbooks on system theory (e.g. Ellis (2012)).

The set of experiments conducted in this thesis can provide great applied benefits to the field of vehicle motion simulation, where knowledge on self-motion perception is sought to improve algorithms controlling simulator motion. As described in section 1.4.3, knowing how the physical motion is (or is not) perceived allow for computing simulator motions that preserve the correct self-motion perception even when the physical motion is not faithfully reproduced. Results from the experiments presented in chapter 6 suggest that humans are insensitive to roll rates slower than ~3.3 deg/s while simulating a passive negotiation of a curve. This result confirms previous findings on pitch rate sensitivity (Groen and Bles 2004) and provides experimental support for the tuning of the tilt-coordination channel of common washout filters. Due to the physical and mental demand of an active task, a further decrease in self-motion sensitivity is expected when drivers are in control of the vehicle (Hosman and van der Vaart 1978; Nesti et al. 2012), suggesting that tilt-coordination can exploit faster tilts to further expand the perceptual workspace of the simulator. Results presented in chapter 6, however, advise caution in concluding that an active task impairs motion sensitivity, as such an effect was only observed in a fraction of the participants. Motion cueing algorithms alternative to the classic washout filters are recently being proposed (Telban et al. 2005; Beghi et al. 2012). One interesting idea is to better exploit the large and growing body of experimental studies on self-motion perception by including perception models of self-motion in their framework. The experiments conducted within this thesis are of great relevance for the development of such algorithms, particularly because of the employed variety of motion stimuli closely resembling the complexity of a vehicle motion.

1.6 DECLARATION OF CONTRIBUTION

The present thesis comprises a collection of manuscripts that, at the time of thesis submission, are either published or prepared for publication in scientific journals. Bibliographic details about the manuscripts are reported below and the chapters where they appear are indicated, together with a description of the contribution of each author.

Nesti A, Beykirch KA, MacNeilage PR, Barnett-Cowan M, Bülthoff HH (2014) The importance of stimulus noise analysis for self-motion studies. PLoS ONE vol 9(4): e94570. [Chapter 2]:

> The idea for the study was proposed by the candidate. The experimental design, software, stimulus generation and data collection were predominantly developed and finalized by the candidate. The co-authors Beykirch, MacNeilage, Barnett-Cowan and Bülthoff supervised the candidates by advising, sharing knowledge, providing comments and criticisms and revising the manuscripts.

Nesti A, Barnett-Cowan M, Macneilage PR, Bülthoff HH (2014) Human sensitivity to vertical self-motion. Experimental Brain Research vol 232, pp. 303–314. [Chapter 3]:

> The idea for the study was proposed by the candidate. The experimental design, software, stimulus generation and data collection were predominantly developed and finalized by the candidate. The co-authors Barnett-Cowan, MacNeilage and Bülthoff supervised the candidates by advising, sharing knowledge, providing comments and criticisms and revising the manuscripts.

Nesti A, Beykirch KA, Pretto P, Bülthoff HH (2015) Self-motion sensitivity to visual yaw rotations in humans. Experimental Brain Research vol 233(3), pp 861-899 [Chapter 4]:

> The idea for the study was proposed by the candidate. The experimental design, software, stimulus generation and data collection were predominantly developed and finalized by the candidate. The co-authors Beykirch, Pretto and Bülthoff supervised the candidates by advising, sharing knowledge, providing comments and criticisms and revising the manuscripts.

Nesti A, Beykirch KA, Pretto P, Bülthoff HH. Human sensitivity to head-centred visual-inertial yaw rotations. Experimental Brain Research (submitted) [Chapter 5]:

> The idea for the study was proposed by the candidate. The experimental design, software, stimulus generation and data collection were predominantly developed and finalized by the candidate. The co-authors Beykirch, Pretto and Bülthoff supervised the candidates by advising, sharing knowledge, providing comments and criticisms and revising the manuscripts.

Nesti A, Nooij S, Losert M, Bülthoff HH, Pretto P. Variable roll-rate perception in driving simulation. Transactions of the Society for Modeling and Simulation International (submitted) [Chapter 6]:

> The idea for the study was proposed by the candidate. The experimental design, software, stimulus generation and data collection were predominantly developed and finalized by the candidate. The co-authors Nooij, Losert, Bülthoff and Pretto supervised the candidates by advising, sharing knowledge, providing comments and criticisms and revising the manuscripts.

Parts of this thesis work were also presented at the following conferences:

- Nesti A, Barnett-Cowan M, MacNeilage P, Bülthoff HH (2012) Differential Thresholds for Vertical Motion, 22nd Okulomotoriktreffen Zürich-München (ZüMü 2012), Zürich, CH.

- Nesti A, Barnett-Cowan M, Bülthoff HH, Pretto P (2012) Roll rate thresholds in driving simulation, 13th International Multisensory Research Forum (IMRF 2012), Oxford, UK.

- Nesti A, Masone C, Barnett-Cowan M, Robuffo Giordano P, Bülthoff HH, Pretto P (2012) Roll rate thresholds and perceived realism in driving simulation. Driving simulation conference, Paris.

- Nesti A, Beykirch K, Pretto P, Bülthoff HH (November-2012) Human sensitivity to different motion intensities, 13th Conference of the Junior Neuroscientists of Tübingen (NeNA 2012), Schramberg, DE.

- Nesti A, Beykirch K, Pretto P, Bülthoff HH (2014) Human sensitivity to visual-inertial self-motion, Neural Control of Movements Society meeting (NCM 2014), Amsterdam, NL.

- Nesti A, Beykirch K , Pretto P and Bülthoff HH (2014) Human self-motion sensitivity to visual yaw rotations, Vision Science Society meeting (VSS 2014), St Pete, Florida, USA.

- Pretto P, Nesti A, Nooij S, Losert M and Bülthoff HH (2014) Variable roll rate perception in driving simulation, Vision Science Society meeting (VSS 2014), St Pete, Florida, USA.

- Pretto P, Nesti A, Nooij S, Losert M and Bülthoff HH. (2014) Variable roll-rate perception in driving simulation. Driving simulation conference, Paris.

2

THE IMPORTANCE OF STIMULUS NOISE ANALYSIS FOR SELF-MOTION STUDIES

This chapter has been reproduced from an article published in PloS ONE: Nesti A, Beykirch KA, MacNeilage PR, Barnett-Cowan M, Bülthoff HH (2014) The importance of stimulus noise analysis for self-motion studies. PLoS ONE vol 9(4): e94570.

2.1 ABSTRACT

Motion simulators are widely employed in basic and applied research to study the neural mechanisms of perception and action during inertial stimulation. In these studies, uncontrolled simulator-introduced noise inevitably leads to a disparity between the reproduced motion and the trajectories meticulously designed by the experimenter, possibly resulting in undesired motion cues to the investigated system. Understanding actual simulator responses to different motion commands is therefore a crucial yet often underestimated step towards the interpretation of experimental results. In this work, we developed analysis methods based on signal processing techniques to quantify the noise in the actual motion, and its deterministic and stochastic components. Our methods allow comparisons between commanded and actual motion as well as between different actual motion profiles. A specific practical example from one of our studies is used to illustrate the methodologies and their relevance, but this does not detract from its general applicability. Analyses of the simulator's inertial recordings show direction-dependent noise and nonlinearity related to the command amplitude. The Signal-to-Noise Ratio is one order of magnitude higher for the larger motion amplitudes we tested, compared to the smaller motion amplitudes. Simulator-introduced noise is found to be primarily of deterministic nature, particularly for the stronger motion intensities. The effect of simulator noise on quantification of animal/human motion sensitivity is discussed. We conclude that accurate recording and characterization of executed simulator motion are a crucial prerequisite for the investigation of uncertainty in self-motion perception.

2.2 INTRODUCTION

For more than a century, motion simulators have been employed in neurophysiological, psychophysical and behavioural studies that aim to inform the neural and cognitive processes of self-motion perception (Mach 1875; Bárány 1907; Lowenstein 1955; Fernández and Goldberg 1976a; Benson et al. 1989; Baloh et al. 2011; Bertolini et al. 2012; Lopez et al. 2013), as well as predicting human behaviours such as balance or aircraft control (Hosman and van der Vaart 1978; Richerson et al. 2003; Mouchnino and Blouin 2013). In all these studies, motion trajectories executed by the simulator inevitably deviate from the commanded motion. This deviation is due to the mechanics of the device and results in motion distortions that affect amplitudes, frequencies and phases of the commanded trajectories (AGARD 1979). Throughout this chapter we define total noise as the components of the actual motion that are not present in the commanded motion. We further define the total noise as the sum of a deterministic component, reproducible across repetitions of the same trajectory (e.g. mechanical deformations due to the inertia of the simulator), and a stochastic component, representing the random component of the total noise. All sensors, including the human self-motion sensory systems (visual, vestibular, auditory and somatosensory), are frequency dependent (see for example Fernández and Goldberg (1976a), Ellis (2012)), and signal processing performance can be directly affected by the level of total noise in the system. Moreover, the simulator noise can provide indirect self-motion cues such as velocity dependent vibrations (Seidman 2008). Response measurements such as neural, perceptual, eye movement or balance recordings should therefore be analysed with a sound understanding of the simulator capabilities and limitations, as well as the impact these limitations have on the results, so as to avoid erroneous interpretations of the data.

While there has been significant prior research on designing, diagnosing and comparing motion systems (AGARD 1979), these methods are not often employed by neuroscientists to dissociate simulator noise from physiological noise in the interpretation of neurophysiological, physiological and behavioural responses. Only a few studies on human self-motion perception address the issue of simulator-introduced noise by recording the actual motion produced. The analyses presented in these perceptual studies can be graphical and/or statistical. In a graphical analysis (see for example Naseri and Grant (2012), Roditi and Crane (2012)), a graphical representation of the simulator's capability is provided by plotting (in the time and/or frequency domain) the different motion recordings together. A statistical analysis, on the other hand, objectively compares recordings of different motion profiles by applying statistical tests. Note that both statistical and graphical analyses can be used to compare either two different actual motions or commanded versus actual motion. Here we summarize the main statistical approaches used so far to assess the influence of simulator noise on perceptual thresholds for self-motion and we present new methodologies for the noise analysis. These methodologies take inspiration from classic techniques for measuring the dynamic qualities of motion simulators (AGARD 1979) and adapt them where necessary to better dissociate between simulator and physiological noise.

To facilitate the description of the methodologies and their relevance, we will use a psychophysical study conducted by the authors (Nesti et al. 2014a). Briefly, a motion simulator was used to investigate human sensitivity to linear vertical self-motion in a range of 0 - 2 m/s^2. Participants were asked to discriminate a reference motion, repeated unchanged for every trial, from a comparison motion, iteratively adjusted in amplitude to measure the participants' motion discrimination thresholds. Different reference motions were tested in different experimental conditions. When interpreting the experimental results, the undesired noise introduced by the simulator is of concern for two main reasons:

1. The total noise level of reference and comparison motions within each condition, if noticeably different, would provide additional cues to the participants.

2. The increase of the total noise level with motion intensity, if non-linear, would lead to a non-constant Signal-to-Noise Ratio (SNR), resulting in differences in stimulus quality across the tested motion range.

Note that, although different experimental procedures have been proposed and used in the literature to investigate the perception of self-motion, none are immune to these problems.

The first point has been raised already by MacNeilage et al. (2010b) and by Mallery et al. (2010). In each of these studies, motions recorded from an inertial measurement unit (IMU) at different commanded amplitudes were analysed to assess their role in the experiments. In MacNeilage et al. (2010b) each of their commanded trajectories was recorded multiple times (13 to 19 repetitions) and the averaged signal was subtracted from each trace to isolate the stochastic noise. Note that averaging over many repetitions causes the stochastic noise to decrease with the square root of the number of trials averaged (Van Drongelen 2008), whereas the deterministic component is always present in the average signal no matter how many trials are averaged. The Fourier transform of each trace was then computed to obtain the amplitude-frequency spectrum of the stochastic noise. An ANOVA of the spectra (0.5 to 100 Hz in 0.5 Hz increments) showed no significant differences between profiles. This method provides an objective way to quantify the amount of stochastic noise in each profile by looking at the amplitude spectrum of the frequencies after the average signal is removed. Note that this procedure not only removes the commanded motion signal but also any deterministic component of the total noise. However, if there is reason to believe that the deterministic noise also depends on the motion intensity (e.g. if the amplitude of the deterministic noise increases with the amplitude of the command), the deterministic noise should not be excluded from the motion analysis, as it can provide a noticeable cue.

A different approach, employed by Mallery et al. (2010), suggests comparing two different stimuli by treating the two digital IMU measures as two different distributions after the commanded signal is filtered out in the frequency domain. A t-test between these two distributions is used to show that the amount of total noise is not significantly different. Because the t-test is specifically designed to compare the means of two populations, this method is able to detect changes in the total noise mean but remains insensitive to changes in the total noise amplitude (the distribution extremes) as long as the two signals have similar means. It is however reasonable to expect that the end-effector of the simulator oscillates around the desired trajectory yielding mean simulator noise close to zero for every trajectory. On the other hand, any effect of motion intensity on the amplitude of the noise will not be detected. For this reason, we did not apply this methodology in the present work.

To the best of our knowledge, the second point, concerning changes in the signal quality across the tested motion intensity range, has never been addressed in any psychophysical study on self-motion perception. Substantial evidence indicates that the SNR of motion simulators depends on the commanded motion intensity and frequency (cf. AGARD (1979), Grant et al. (2001)) and that human self-motion sensitivity is affected by stimulus SNR (Greig 1988). Therefore, we believe that signal quality is a potential confound in the analysis of self-motion responses and should always be given careful consideration. It is not our goal here to investigate the effect of the motion SNR on human self-motion sensitivity. Instead we present an SNR analysis of the motion profiles, which constitutes an essential step for a correct interpretation of experimental results.

For our chosen example study (Nesti et al. 2014a), it is most appropriate to analyse the total noise because each trial consisted of both a reference and comparison motion, of unequal amplitude, leading to potential differences in both deterministic and stochastic noise. Using total noise is best when the deterministic component may alter the results. However, for comparison, here we also present methodologies to quantify the relative contribution of stochastic and deterministic components and their dependencies on the commanded motion. Separate analysis of deterministic and stochastic components is relevant, for example, in studies where many repetitions of the same command are employed (e.g. for measuring gains of the vestibular ocular reflex). In these cases the actual motion stimulus is the motion command combined with the deterministic noise, and deviations of eye traces from the actual motion command are caused by physiological noise (focus of interest) and stochastic noise (undesired simulator-introduced variability).

2.3 METHODS

The study was conducted using the Max Planck Institute CyberMotion Simulator, a 6-degrees-of-freedom anthropomorphic robot-arm, able to provide a large variety of motion stimuli, with a maximal vertical displacement of about 1.4 m and a maximal vertical linear acceleration of about 5 m/s^2 (for technical details refer to Robocoaster, KUKA Roboter GmbH, Germany; Teufel et al. 2007; Barnett-Cowan et al. 2012). IMU traces were acquired for 10 reference stimuli (1 Hz sinusoidal acceleration profiles with peak amplitudes of 0.07, 0.3, 1.1, 1.6 and 2 m/s^2, both upward and downward) with a 3D accelerometer (YEI 3-Space Sensor, 500 Hz) attached rigidly on the back of the simulator seat. While our trajectories did not involve rotations, it is important to note that for rotational trajectories, seat and head motions differ and placing the IMU on the participant's head is a more sensible choice. For each reference stimulus, we additionally recorded two comparison stimuli whose peak intensity was raised (higher comparison) and lowered (lower comparison) by two corresponding discrimination thresholds, so as to quantify the noise level changes within stimuli of the same condition. The discrimination thresholds associated with the reference stimuli are 0.02 (unpublished observation), 0.09, 0.21, 0.23 and 0.25 m/s^2, respectively (Nesti et al. 2014a). Each profile (Figure 8) was recorded 20 times.

Of the recorded profiles, only the frequency components below 80 Hz were considered for further analyses, under the assumption that for these profiles frequencies higher than 80 Hz do not affect psychophysical performance (in agreement with Mallery et al. (2010)). From each signal the 1 Hz input was subtracted to obtain the total noise signal. Figure 9 illustrates the procedure for a downward acceleration with peak amplitude of 2.5 m/s^2. The deterministic component was obtained by averaging the total noise across repetitions of each profile, and the stochastic component was obtained by subtracting the deterministic component from the total noise.

Two different methods were employed to analyse the total noise level of the profiles: the amplitude-frequency spectrum and the root mean square (rms). These methods are explained in more details in the following sections. Additionally, for the 10 reference stimuli, the SNR was computed to characterize the relationship between the quality of the reproduced motion and the intensity of the commanded motion (section Signal-to-

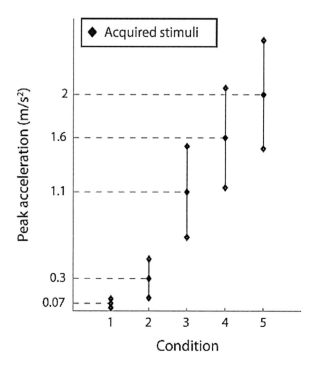

Figure 8 *Graphical representation of peak amplitude for the acquired stimuli for both upward and downward motion. The dashed lines indicate the reference intensities, around which the higher and lower comparison were set.*

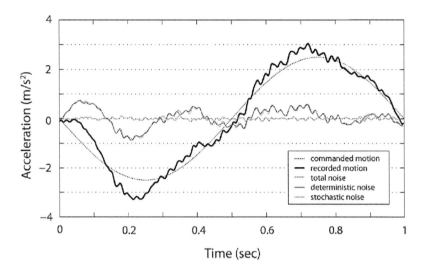

Figure 9 *Example of an acquired motion profile and its component. The total noise (grey dashed line) was obtained by subtracting the input command (black dashed line) from the acquired acceleration profile after low-pass filtering (black thick line). The figure also illustrates the deterministic (black thin line) and stochastic (grey thin line) components of the total noise of the recorded profile.*

noise ratio analysis). We further analyse these stimuli in terms of the deterministic and stochastic components of their total noise (section Deterministic and stochastic noise analysis). Signal processing and statistical analysis were performed in MATLAB (2012a) using custom-written code and the Statistics toolbox.

Amplitude-frequency spectrum analysis

The total noise affecting the motion profiles can be objectively quantified by its amplitude-frequency spectrum. Such an indicator has the advantage of providing details about which frequencies are more affected by the noise. This approach, based on MacNeilage et al. (2010b), differs from the original work described previously since from each acquired trace only the input command is removed, rather than the average over several repetitions (which contains both input and deterministic noise). This allows for an analysis of the total rather than the stochastic noise. Force/exponential windows of one and two seconds, respectively, were applied to the original signals according to equation 1. This allows for reduction of frequency leakage (Halvorsen and Brown 1977; McConnell and Varoto 1995) without altering the amplitude of the total noise signal contained in the first second of the window.

$$\hat{x}_i = \begin{cases} x_i & if\ 1 \leq i < 500 \\ x_i * \exp\left(-\dfrac{10}{N-1} * (i - 500)\right) & if\ 500 \leq i < N \end{cases} \tag{1}$$

where $\hat{x}_i$ is the i-th sample of the windowed signal, x_i is the i-th sampled measure of noise and N is the number of samples (in this case 1500 samples).

After Fourier transforming the windowed signals we obtained 3 groups of 20 amplitude-frequency spectra for each condition (Figure 8): one group for the reference motion, one for the higher comparison and one for the lower comparison. A typical amplitude-frequency spectrum is presented in Figure 10. Note that it is possible to infer the main frequencies that compose the total noise from spectral analysis. The 20 amplitude spectra of each reference motion were tested against the 20 amplitude spectra of their corresponding higher and lower comparisons independently by using an ANOVA with 2 factors: frequency (0 to 80 Hz in 0.33-Hz increments, yielding 241 values) and motion profile (reference or comparison).

The same analysis was also performed on the stochastic noise so as to allow for comparison with the *analysis of vibration* reported in MacNeilage et al. (2010b) (note the change in terminology from "vibration" to "stochastic noise"). Results of these analyses are shown in Table 2.

Root mean square analysis

The rms, or quadratic mean, is a measure of the magnitude of a varying quantity. Here, its discrete formula is used to objectively quantify the noise level of each 1 sec signal:

$$rms = \sqrt{\frac{1}{N}\sum_{i=1}^{N} x_i^2} \qquad (2)$$

where x_i is the i-th sampled measure of noise and N is the number of samples (in this case 500 samples). For each condition we obtained 3 groups of 20 rms values each: one group for the reference motion, one for the higher comparison and one for the lower comparison, which were all repeated 20 times. To determine whether noise level changes within conditions are reliable cues for motion amplitude discrimination, every reference rms group was tested for statistically significant differences (unpaired 2-sample t-test) against its corresponding higher and lower

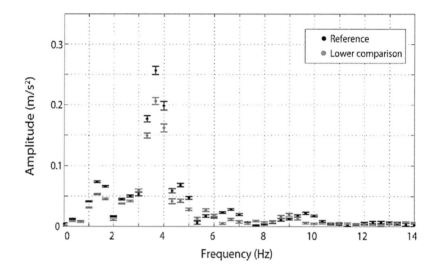

Figure 10 *Example of an amplitude-frequency spectrum for a reference of 1.6 m/s² and corresponding lower comparison. In this example the simulator noise mainly affects frequencies around 4 Hz. Error bars represent standard deviations of the 20 repetitions of each profile. The abscissa is limited to only the first 61 frequency components (0-20 Hz) out of the 241 (0-80 Hz) used for the analysis for better graphical clarity.*

comparison independently. This rms analysis was conducted on the total noise as well as the stochastic component of the noise. Results of these analyses are shown in Table 2.

Signal-to-noise ratio analysis

The SNR is used to express the relative amount of commanded signal and background noise present in each trajectory. A SNR close to 1 indicates that the level of noise in the reproduced motion is comparable to the level of commanded signal. This is often the case for motion simulators when reproducing small accelerations (e.g. <10 cm/s^2 for the simulator tested here) (Zaichik et al. 1999; Grant et al. 2001). Higher SNRs indicate a reproduced motion of higher quality, where the signal level overcomes the noise. We computed the SNR for every repetition of the 10 reference motions according to the following formula:

$$SNR = \left(\frac{rms_{signal}}{rms_{total\_noise}} \right)^2 \tag{3}$$

Differences in the SNRs were tested with an ANOVA with 2 factors: direction (upward or downward) and motion intensity (0.07, 0.3, 1.1, 1.6 and 2 m/s^2). See Figure 11 for results.

Deterministic and stochastic noise analysis

Quantitative measures of the stochastic and deterministic noise components in a reproduced motion allow for characterization of the nature of the noise, perhaps providing important information for deciding how to deal with the noise (see discussion). We calculated the rms of the stochastic and deterministic noise for the 10 reference motions. The Deterministic-to-Stochastic Ratio (DSR) introduced in equation 4 indicates which component is dominant in an analysed profile and the way that the total noise composition changes over different stimulus intensities. The results are presented in Figure 12.

$$DSR = \frac{rms_{\text{det\_noise}}}{rms_{stoc\_noise}} \tag{4}$$

Instrumentation and environmental noise

Even when recording no motion, a certain level of background activity is to be expected in any IMU recording. This is due to electrical interferences as well as specific traits of the IMU, which obviously do not reflect real motion. New software and hardware improvements are continuously being developed to reduce this sensor noise (see e.g. Widrow et al. (1975), Van Drongelen (2008)), however it can never be completely eliminated. The analyses proposed in this manuscript assume that sensor noise, in comparison to motion noise, is negligible. This is often the case for high quality sensors, where sensor shielding strongly reduces electrical interferences. However, in this work the demonstration of the proposed techniques was done using commercial hardware, potentially sensitive to environmental noise such as electrical interferences from the simulator motors. To quantify the level of sensor noise affecting the recordings, IMU data were acquired for approximately 4 minutes while the motion platform was powered but not moving and the rms of the acceleration signal was calculated according to equation 2.

2.4 RESULTS

The tested reference/comparison pairs show significantly different total noise levels both in terms of amplitude-frequency spectrum and rms of the total noise signals in almost every tested pair (Table 2). This indicates that

Compared profiles (reference vs. comparison) [m/s²]	rms analysis [t(38), p values]				Amplitude–frequency spectrum analyses [df = 1, p values]			
	Total Noise		Stochastic Noise		Total Noise		Stochastic Noise	
	Motion Direction		Motion Direction		Motion Direction		Motion Direction	
	up	down	up	down	up	down	up	down
0.07 vs. 0.03	**<0.001**	**<0.001**	0.47	0.68	**<0.001**	**<0.001**	**<0.001**	0.06
0.07 vs. 0.11	**<0.001**	**<0.001**	0.66	0.11	**<0.001**	**<0.001**	**0.03**	**<0.001**
0.3 vs. 0.12	**<0.001**	**<0.001**	0.05	**0.02**	**<0.001**	**<0.001**	**<0.001**	**<0.001**
0.3 vs. 0.48	**<0.001**	**<0.001**	0.18	**0.002**	**<0.001**	**<0.001**	**0.02**	**<0.001**
1.1 vs. 0.68	**<0.001**	**<0.001**	0.39	0.57	**<0.001**	**<0.001**	**<0.001**	**<0.001**
1.1 vs. 1.52	**<0.001**	**<0.001**	0.98	0.38	**<0.001**	**<0.001**	0.11	**0.002**
1.6 vs. 1.14	**<0.001**	**0.005**	0.36	0.06	**<0.001**	**<0.001**	**<0.001**	**<0.001**
1.6 vs. 2.06	**<0.001**	**<0.001**	0.38	0.19	**<0.001**	**<0.001**	**<0.001**	**<0.001**
2 vs. 1.5	**<0.001**	**<0.001**	0.08	**<0.001**	**<0.001**	**<0.001**	**<0.001**	**<0.001**
2 vs. 2.5	0.12	**<0.001**	0.21	**0.01**	**<0.001**	**<0.001**	**<0.001**	**<0.001**

Table 2 *P-values resulting from the two analyses comparing the levels of total and stochastic noise around each reference for both upward (up) and downward (down) movements. Effects with a p-value <0.05 are considered as significant and appear in bold.*

the total noise introduced by the simulator depends on the commanded motion intensity.

As expected, the stochastic noise (quantified by its rms value) correlates with the inverse of the square root of the number of trials averaged (in all groups $0.77 \leq r \leq 0.99$, average $r = 0.89$). However, sensor noise analysis revealed comparable noise levels between the no-motion profile (rms mean +/- std: = 0.05 +/- 0.008 m/s^2) and the motion profiles (see Figure 12A). This suggests that stochastic noise is likely to reflect predominantly sensor noise rather than real motion noise, such that the level of stochastic noise in the recorded trajectories is impossible to resolve with the current equipment. Results from the amplitude-frequency spectrum analysis and the rms analysis performed on the stochastic noise alone (see Table 2) should therefore be interpreted with caution, as they likely reflect features of the sensor noise rather than physical motion. The overall level of deterministic noise in the motion trajectories (see Figure 12A) is instead often higher than the sensor noise rms, making it unlikely for the sensor noise to significantly influence the analyses of the deterministic and total noise components.

The total noise rms was found to increase non-linearly with the amplitude of the commanded reference signal ($F(4,199)=25037$, $p<0.001$) over the tested range, which leads to SNRs that depend on the motion intensity (Figure 11). The results show SNRs one order of magnitude higher for the stronger than for the weaker measured profiles. Moreover, SNRs were overall better for downward compared to upward motion ($F(1,199)=1560$, $p<0.001$). These results suggest that for perceptual systems whose sensitivity increases with SNRs, regardless of motion intensity, motion discrimination using this particular simulator should be proportionally better for downward as compared to upward motions and for higher as compared to lower motion intensities. Experimental data on human motion sensitivity over wide motion ranges (Zaichik et al. 1999; Mallery et al. 2010; Naseri and Grant 2012; Nesti et al. 2014a), however, do not show such

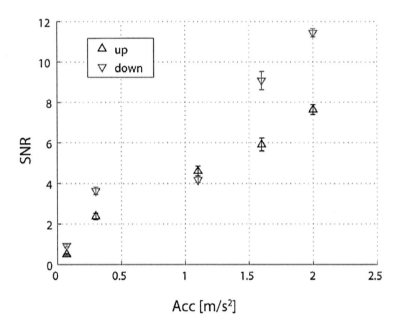

Figure 11 *The SNR for upward and downward motion profiles increases as a function of motion intensity. Error bars represent standard deviations of the 20 repetitions of each profile.*

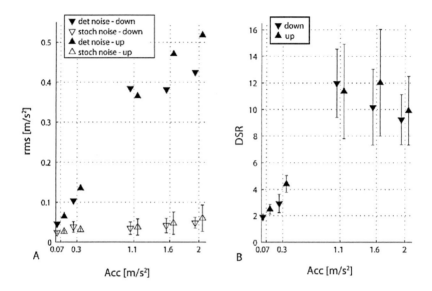

Figure 12 *Relative contributions of deterministic and stochastic components to the total noise.* **Panel A:** *Each reference stimulus recording is associated with the rms of its deterministic (filled triangles) and stochastic component (empty triangles) for both upward (upward pointing triangles) and downward (downward pointing triangles) motions.* **Panel B:** *The DSR for upward and downward profiles. Both panels indicate a predominance of deterministic over stochastic noise in the recorded profiles. Error bars represent standard deviations of the 20 repetitions of each profile.*

behaviour, suggesting an additional noise source inherent to the perceptual system which is proportional to stimulus intensity. Asymmetries in vertical motion sensitivity (Nesti et al. 2014a), instead, might be entirely explained by the results of the SNR analysis.

The rms of the stochastic and deterministic noise of each reference profile for upward and downward movements is presented in Figure 12A. The level of the deterministic noise increases notably with the stimulus intensity and overall is higher than the level of the stochastic noise, which on the other hand remains rather constant over the tested motion range. Consequently, DSRs increase for stronger motion intensities. Note that, as stated above, the stochastic components of the motion noise are likely lower than the noise of the employed sensor. Therefore, the rms of the stochastic noise represents an "upper bound" for the true rms value of the stochastic noise in the motion profiles. Nevertheless, from these results emerges a dominance of the deterministic component of the total noise over the stochastic component, particularly at the higher motion intensities. This suggests that deterministic noise is more likely than stochastic noise to impact self-motion perception in this experimental paradigm.

2.5 DISCUSSION

The analyses presented here allow characterization of the noise introduced by the simulator when reproducing commanded trajectories. They provide sensitive methods to compare the noise level of different commanded stimuli and to graphically and statistically describe the reproduced motion. Results show that the total noise of the simulator increases with the amplitude of the command in a nonlinear way, leading to an SNR that increases with the motion intensity. Even for relatively small changes in the amplitude of the commanded motion, changes in the measured noise are statistically significant. This raises the question of whether a human, when asked to report changes in the motion intensity, could use changes in the simulator noise as a cue rather than changes in the signal itself.

It is reasonable to assume that, if our analyses of the IMU signals do not demonstrate these differences, the human will also not detect them. It is however erroneous to conclude the opposite. The simulator motion is available to the CNS only after being processed by the sensory systems (vestibular, somatosensory and proprioceptive), whose dynamics are imperfect due to frequency dependences and noise (Gong and Merfeld 2002; Tahboub and Mergner 2007). Furthermore, the way that the CNS deals with these signals is likely different from the statistical analysis employed here. Consider as an example the mp3 and AAC encoding techniques in music: even though the frequency spectra of the original and compressed signals look dramatically different, they are virtually indistinguishable to a human observer due to the inability of the auditory system to perceive the differences (Meares et al. 1998). Although the frequency response range of the otolith organs of the vestibular system is estimated to be between 0 Hz and 1.6 Hz (Grant and Cotton 1991), the contribution of the other sensory systems should not be neglected. For this reason we did not filter the data with a model of the vestibular system.

Given these results and the previous considerations, speculations can be made as to how simulator noise affects human perception of motion intensity. To distinguish between motions at different intensities, differences in the neuronal signals that reach the CNS need to overcome the internal noise level (Goldberg 2000; Sadeghi et al. 2007; Yu et al. 2012). If the internal noise is small relative to the total noise of the motion, motion stimuli with high SNR are likely to generate neuronal signals that also have a high SNR. This would facilitate the process of detecting changes in the motion intensity. Additionally, human self-motion sensitivity could be enhanced by changes in the motion noise level if those changes are captured by the human sensors and successfully processed by the CNS. An accurate analysis of the noise of the experimental setup is therefore of great importance for the active research field investigating the noise in the nervous system and its effect on information processing (Faisal et al. 2008).

By including a demonstration of the proposed methodologies in this chapter, we raise the practical concern of sensor noise, which affects any measuring setup regardless of the sensor nature (e.g. IMUs, optical trackers, etc.). Sensor measures during no motion allowed us to estimate the level of sensor noise and to conclude that stochastic components of the noise motion are likely smaller than the level of sensor noise, indicating that simulator-introduced noise is primarily of deterministic nature. A more precise quantification of the stochastic motion noise was precluded by the stochastic sensor noise. Overall, caution is advised in the interpretation of sensor measurements as much as in the interpretation of responses to noisy self-motion stimuli.

To address the stochastic and deterministic composition of the total noise, we have provided their formal definitions and a methodology for extracting them from IMU recordings. Other than using the DSR introduced here, deterministic and stochastic noise can also be compared using a frequency analysis, which is particularly useful for highlighting the frequency ranges affected by the two types of noise (see the examples in Figure 13).

Whenever possible effort should be spent in minimizing the deterministic noise, so that its impact during experiments is also minimized. This is particularly beneficial in cases where a DSR analysis indicates a predominantly deterministic nature of the noise. Deterministic noise can be reduced by using iterative learning control algorithms (Grant et al. 2007; Ahn et al. 2007): given a desired trajectory these algorithms iteratively process IMU recordings of the simulator motion and modify the simulator commands so as to track the desired trajectory as closely as possible.

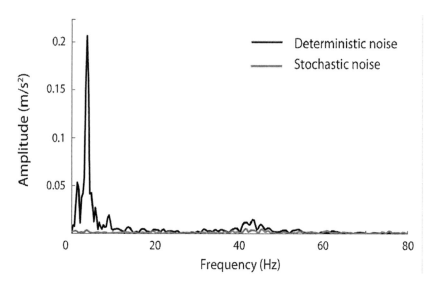

Figure 13 *Amplitude-frequency spectrum of the deterministic and stochastic noise components of the acceleration profile whose total noise is illustrated in Figure 9. The DSR of this profile is 13.48.*

2.6 CONCLUSION

Simulator-introduced noise is a recurrent concern for neuroscientists who use motion simulators to investigate the neural and cognitive mechanisms of self-motion perception. In this work we developed straightforward graphical and statistical techniques for the analysis of motion stimuli commonly employed in self-motion studies. Rather than measuring the dynamic qualities of a motion system, these analyses allow for dissociation between simulator and physiological noise and therefore constitute a valuable set of tools for the interpretation of neurophysiological and behavioural responses, as well as for meaningful comparisons across the existing literature. We further illustrated these analyses and their relevance using a prior study on human self-motion perception. Results clearly demonstrate the importance of noise, including both stochastic and deterministic components. It should be noted that, although the methods are of general application, the presented results hold for the employed simulator only and other simulators are expected to show substantial differences in their dynamic responses.

ACKNOWLEDGMENTS

We gratefully thank Michael Kerger and Harald Teufel for technical assistance.

GRANTS

AN, KAB and MB-C were supported by funds from the Max Planck Society. PRM was supported by the German Federal Ministry of Education and Research under the Grant code 01 EO 0901. This work was also supported by the Brain Korea 21 PLUS Program through the National Research Foundation of Korea funded by the Ministry of Education. The funders had no role in study design, data collection and analysis, decision to publish, or preparation of the manuscript.

COPYRIGHT

3

HUMAN SENSITIVITY TO VERTICAL SELF-MOTION

This chapter has been reproduced from an article published in Experimental Brain Research: Nesti A, Barnett-Cowan M, Macneilage PR, Bülthoff HH (2014a) Human sensitivity to vertical self-motion. Exp Brain Res vol 232 pp 303–314.

3.1 ABSTRACT

Perceiving vertical self-motion is crucial for maintaining balance as well as for controlling an aircraft. Whereas heave absolute thresholds have been exhaustively studied, little work has been done in investigating how vertical sensitivity depends on motion intensity (i.e. differential thresholds). Here we measure human sensitivity for 1 Hz sinusoidal accelerations for 10 participants in darkness. Absolute and differential thresholds are measured for upward and downward translations independently at 5 different peak amplitudes ranging from 0 to 2 m/s^2. Overall vertical differential thresholds are higher than horizontal differential thresholds found in the literature. Psychometric functions are fit in linear and logarithmic space, with goodness of fit being similar in both cases. Differential thresholds are higher for upward as compared to downward motion and increase with stimulus intensity following a trend best described by two power laws. The power laws' exponents of 0.60 and 0.42 for upward and downward motion respectively deviate from Weber's Law in that thresholds increase less than expected at high stimulus intensity. We speculate that increased sensitivity at high accelerations and greater sensitivity to downward than upward self-motion may reflect adaptations to avoid falling.

3.2 INTRODUCTION

Humans can move in three dimensions. Compared to fore-aft and lateral movements (also called surge and sway respectively), vertical self-motion (or heave) is particularly physically and ecologically constrained due to the constant force of gravity. The downward force of gravity is equivalent to an upward acceleration that is detected by the otolith organs of the vestibular system. At rest, the otoliths indicate an upward acceleration at the rate of 9.8 m/sec^2. But when at rest we do not feel as though we are moving, suggesting that the brain must compensate for gravity's constant influence. Despite this compensation, the presence of a non-zero pedestal stimulus could lead to asymmetric sensitivity to earth-vertical motion. In particular, it is not yet known whether sensitivities to upward and downward self-motion differ. Here we investigate this question by measuring how sensitivity depends on the magnitude and direction of vertical acceleration.

Human self-motion perception arises from central processing of sensory information from the visual, vestibular, auditory and somatosensory systems. The first step to characterize this complex perceptual process is to understand how these various sources individually contribute to the subjective representation of physical motion. Such characterization finds immediate application in the field of vehicle simulation and in clinical assessment of balance disorders. Motion algorithms for dynamic simulators rely on human self-motion perception knowledge to provide, within their limited workspace, the most realistic motion sensation. Predicting the perception of vertical self-motion is especially important for flight simulation, particularly during takeoff and landing manoeuvres. In the medical field, current protocols for diagnosing orientation perception disorders rely on measuring oculomotor reflexes (Bárány 1921; Halmagyi and Curthoys 1988). However, perception and reflexes do not always correlate (Kanayama et al. 1995; Merfeld et al. 2005a; Merfeld et al. 2005b). Psychophysical measurements are therefore a helpful tool for localizing the disorder source (Merfeld et al. 2010; Agrawal et al. 2013).

A common method for investigating human self-motion perception is to measure perceptual thresholds by asking participants to make judgments based on the provided motion stimulation. These experimental studies can be divided into two main categories: estimation of absolute thresholds (the smallest detectable *level* of a stimulus intensity) and estimation of differential thresholds (the smallest detectable *change* in stimulus intensity). To measure absolute thresholds participants usually perform either a detection task (report the presence of motion) or a direction discrimination task (report motion direction, sometimes also referred to as direction recognition task). To measure differential thresholds participants perform an amplitude discrimination task (discriminate between two different movements).

Whereas detection and direction discrimination thresholds for human self-motion have been exhaustively studied, little work has been done in investigating our ability to discriminate vertical self-movements over different ranges of motion intensities. Human differential thresholds to linear self-motion have been investigated by Naseri and Grant (2012) over a range of $0.5 - 2$ m/s^2 for surge motion and by Zaichik et al. (1999) over a range of $0 - 0.6$ m/s^2 for surge, sway and heave motion. Their works show that differential thresholds increase with stimulus intensity following Weber's perceptual law (Fechner 1860). According to this fundamental law of psychophysics, the change in a stimulus that is just noticeable (differential threshold) is a constant ratio of the original stimulus. However, this was not confirmed by MacNeilage et al. (2010a), who found no significant change in sensitivity for surge and heave motion over a range of

$0 - 0.3$ m/s^2. Thus, it remains unresolved whether perceived linear self-motion (i) is independent from stimulus intensity; (ii) follows Weber's law; (iii) follows a different (nonlinear) law. To assess these competing hypotheses, the present work sets out to describe differential thresholds for heave motion in the absence of visual cues as a function of motion intensity.

Assessing whether or not perceived self-motion follows a Weber's law is important as many current models used to mathematically describe the process of self-motion perception (Borah et al. 1988; Bos and Bles 2002; Zupan et al. 2002) assume that sensitivity to supra-threshold self-motion is not affected by motion intensity once absolute threshold is overcome. This is true for the vestibular ocular reflex (VOR), which maintains a constant level of accuracy and precision over a wide range of motion intensities (Pulaski et al. 1981; Zanker 1995). The VOR, elicited by the vestibular system in response to head-in-space motion, is one of the vestibular mechanisms that maintain gaze and postural stability and, like self-motion perception, may be modulated from higher-level neural processes. However vestibular perception and action employ qualitatively different mechanisms (Barnett-Cowan et al. 2005; Merfeld et al. 2005a; Merfeld et al. 2005b; Bertolini et al. 2011). If psychophysical evidence indicates a nonlinear relationship between self-motion perception and stimulus intensity, the accuracy of self-motion perception models over a wide motion range would benefit greatly from the implementation of differential thresholds.

With this work we also address the question of asymmetries in motion perception between upward and downward movements. Evidence of asymmetry in vertical self-motion perceptual thresholds was previously reported by Benson et al. (1986) for participants lying on the back, who measured perceptual thresholds for vertical translations in head coordinates. They observed that movements in the footward direction are more readily perceived than movements in the headward direction. Notice that for participants lying on the back the effect of gravity on head-relative heave motion is symmetric, suggesting that gravity is not responsible for vertical perceptual asymmetries. However, these results conflict with earlier (Melvill Jones and Young 1978) and more recent (MacNeilage et al. 2010a; Roditi and Crane 2012) reports, where no significant differences are reported between opposite direction movements along the vertical axis. Determining whether discrepancies in vertical self-motion sensitivity exist is not only important to consider when designing motion control algorithms, such a discrepancy would yield insight into the dynamics of the inertial sensors and/or central processing of self-motion information. Asymmetries in the perception of upward and downward motion would reflect high-level processes in the central neural system (e.g. higher sensitivity to downward movements to prevent falls). Alternatively, similar sensitivities would suggest that human self-motion perception perfectly compensates for gravity both at rest and during motion. In this work upward and downward sensitivity will be compared not only for movements close to absolute threshold but over a wide range of motion.

3.3 METHODS

Participants

Ten subjects (3 females; aged 20-31 years), nine naïve and one author (MB-C), participated in the study and gave their informed written consent in accordance with the ethical standards specified by the 1964 Declaration of Helsinki prior to their inclusion in the study. Participants reported having no vestibular or other neurological disorders and no susceptibility to motion sickness.

Setup

The experiment was conducted using the Max-Planck Institute CyberMotion Simulator. This motion simulator is based on a 6-degrees-of-freedom anthropomorphic robot-arm and can provide a large variety of motion stimuli, with a maximal vertical displacement of about 1.4 m and a maximal linear acceleration of about 5 m/s^2. Further details on its hardware and software specifications are available (Robocoaster, KUKA Roboter GmbH, Germany; Teufel et al. 2007; Barnett-Cowan et al. 2012). Participants were seated in a chair with a 5-point harness (Figure 14). They wore light-proof goggles to eliminate visual information, as well as ear plugs (SNR=33, NRR=29) and headphones with acoustic white noise played back during the movements to eliminate external auditory cues from the simulator motors. None of the participants reported that the motor noise was heard during the experiment. To mask possible air movement cues during the motion, participants wore long trousers and sleeves, and a fan was installed in front of them. The seat and the feet of the participants were covered with foam to mask vibrations of the simulator. The use of a neck brace, combined with careful instruction to maintain an upright position, was assumed to minimize head movements. Head motion was therefore not recorded.

Procedure

Each trial was composed of two consecutive vertical movements in the same direction. One of these movements remained unchanged in every trial (pedestal stimulus), while the second movement (comparison stimulus) systematically varied in amplitude. Accelerations during each interval were single-cycle sinusoidal profiles with 1 sec duration (i.e. 1 Hz, see Figure 15) with amplitudes ranging from 0 to 3.3 m/s^2. This allows for comparison with previous research (Benson et al. 1986; MacNeilage et al. 2010a; Naseri and Grant 2012; Roditi and Crane 2012) and is within the frequency range of flight simulations. Pedestal and comparison stimulus order was randomized to avoid order effects and complications due to motion after-effect (i.e. the influence of a previous motion on the next motion, Crane 2012).

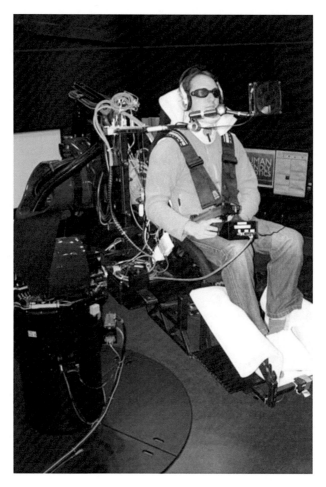

Figure 14 *Experimental setup.*

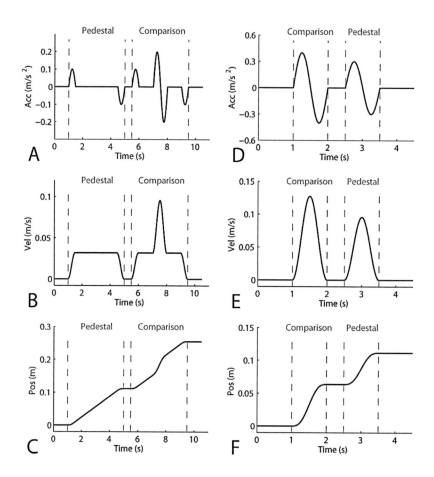

Figure 15 *Acceleration, velocity and position traces for a baseline condition trial (panels A, B and C respectively) and for a trial with 0.3 m/s² pedestal (Panels D, E and F respectively).*

Participants initiated each trial with a button press and, after a 1 sec pause, the movement began. The two movements were separated by a 0.5 sec interval where no motion was provided. After the second movement ended, participants were instructed to indicate via button press "which movement was stronger, the first or second, in terms of highest acceleration, velocity and covered distance". After answering in this two-interval forced-choice task (2IFC) they were moved back to the starting position with a slow, but supra-threshold movement involving all the simulator joints. No feedback was given.

The experiment consisted of five sessions lasting approximately 1.5 hours each and was conducted on separate days. In each session the absolute or differential threshold was assessed for two pedestal amplitudes with separate ~40-minute blocks for upward and downward directions with breaks every 15 minutes to avoid fatigue. Peak pedestal amplitudes were 0 (baseline, see below), 0.3, 1.1, 1.6 or 2 m/s². The 0.3 m/s² pedestal was chosen so to allow comparison with MacNeilage et al. (2010a), while the irregular spacing reflects our interest for highest motion intensities and was obtained with the following formula:

$$p = \frac{ln(1 + x)}{ln(4)} * (2 - 0.3) + 0.3 \qquad (5)$$

Where p is the 4 element vector of pedestal intensities and x is a vector of 4 linearly spaced values between 0 and 3. Each pedestal was tested in the upward and downward direction separately, for a total of 10 conditions per participant.

The two baseline conditions (pedestal of 0 m/s²) employ slightly different methods to measure absolute rather than differential thresholds. In these conditions movements were superimposed on a constant velocity translation in the same direction to avoid additional cues from motor activation (Figure 15A). Velocity was initially increased following a half raised cosine profile for 0.5 secs. Then, after 3 secs of constant velocity motion, velocity was decreased back to zero with the other half of the original raised cosine profile. During the constant-velocity phase, the 1 Hz sinusoidal acceleration profile was superimposed, with a time offset randomly selected from a list of 3 possible values: 0.5, 1 or 1.5 secs. In this condition, participants' were asked to report which of the two intervals had a superimposed acceleration, equivalent to a 2-interval detection task. Note, only 6 of the 10 participants completed this baseline condition. The other 4 participants were not available for further testing.

For each pedestal, 40 comparison stimuli were symmetrically distributed about each pedestal according to equation 6:

$$c = \left[1 \pm \frac{s^2}{\frac{2}{3} * p^2} \right] * p \tag{6}$$

Where c is the vector of comparison stimuli intensities, p is the pedestal stimulus intensity and s is a 21 elements vector linearly spaced between 0 and 2/3 * p. The resulting stimuli are in a range of +/- 67% of the pedestal intensity, with higher stimulus density near the pedestal (Figure 16). In the baseline condition, 21 stimuli were similarly placed between 0 and 0.3 m/s^2.

The percentage of correct answers as a function of motion amplitude was fit with a continuous psychometric function (Kontsevich and Tyler 1999; Tanner 2008). The psychometric function was modelled as a cumulative normal distribution (Figure 17). Two lapse parameters limited to 5% were included into the fit to account for the possibility of accidentally pressing the wrong button even if the direction was correctly perceived. It has been shown that this can significantly improve the fit (Wichmann and Hill 2001). The fitting was performed in logarithmic stimulus space, a choice motivated by the proportional decrease of self-motion sensitivity with

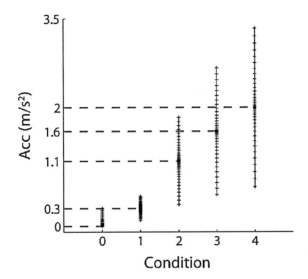

Figure 16 *Graphical representation of the peak amplitude of the comparison stimuli used. The grey dotted lines indicate the pedestal intensities around which the set of stimuli was selected for each condition. Each tick marks a possible acceleration value to be presented.*

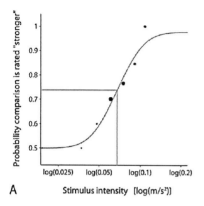
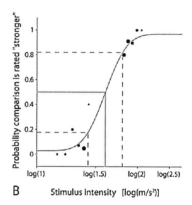

Figure 17 *Psychometric functions fit (grey line) in log stimulus space to the data (black dots) obtained for one participant in the baseline condition (panel A) and for a pedestal of 1.6 m/s^2 (panel B). The probability of rating the comparison stimulus as stronger than the pedestal is on the y axis. The light grey line and the black dashed line represent the mean and the standard deviation of the fitted cumulative Gaussian, respectively.* **Panel A:** *The grey line represents the stimulus that corresponds to the participant's absolute threshold.* **Panel B:** *The black dashed lines represent stimuli that are one standard deviation weaker (left dashed line) or stronger (right dashed line) than the pedestal. According to our definition of differential threshold, the region on the x axis between the black dotted lines encloses stimuli that cannot be distinguished from the pedestal.*

increasing stimulus intensity (Zaichik et al. 1999; Mallery et al. 2010; Naseri and Grant 2012) and because it has already been adopted in previous studies (Soyka et al. 2011; Soyka et al. 2012).

Example psychometric fits are illustrated in Figure 17. In all but the baseline conditions (Figure 17B) the inflection point of the cumulative Gaussian corresponds to chance performance, i.e. 0.5 probability[4] of answering correctly. The standard deviation corresponds to a change in the stimulus amplitude that increases the performance to 0.84 probability of correct identification. This parameter is an indicator of the participant's sensitivity to relative change in motion and is therefore arbitrarily taken as the differential threshold. Normally, the probability of correct discrimination corresponding to a stimulus change of one differential threshold is arbitrarily chosen by the experimenter between 70% and 85% (Mallery et al. 2010; MacNeilage et al. 2010a; Naseri and Grant 2012).

In the baseline condition the chance level of correctly detecting the amplitude discrepancy is at 0.5. Therefore, the inflection point of the cumulative Gaussian is located at 0.5 * (1 − 0.5) + 0.5 = 0.75 (see Figure 17A). The corresponding motion amplitude was considered as the absolute threshold for the perception of linear vertical motion (Soyka et al. 2011; Soyka et al. 2012).

A Bayesian adaptive procedure, based on the method proposed by Kontsevich and Tyler (1999), was used to estimate the psychometric function (Tanner 2008). The basic idea behind the method is to fit a psychometric function to the whole data set after each trial. Simulating the answer of the next trial for each possible comparison stimulus determines which stimulus would minimize the spread of the parameters' posterior distribution according to an entropy-based cost function. Smaller entropies indicate higher confidence that the fitted psychometric function resemble the model underlying the participant's behaviour. The selected stimulus is

---

[4] All the probabilities reported in this paragraph might be slightly affected by the lapse parameters.

considered the most informative and is used in the next trial. Making use of this method allows for a fast and accurate estimation of the psychometric function. In all but the baseline conditions the efficiency of this method is increased by fixing the mean of the psychometric function to the pedestal stimulus value because there is one less free parameter to consider. Note how this allows for the method to select stimuli in regions that are more informative for estimating the standard deviation.

For each condition, participants were tested for at least 100 trials and until the estimate stabilized. Our criterion for stabilization was that the fluctuation in the threshold estimate provided by the Bayesian adaptive method over the last 20 trials (i.e. highest minus lowest value) became smaller than 5% of the highest value. If this criterion was not achieved after a maximum of 200 trials the algorithm stopped and the last estimate was selected as the threshold. No session had to be terminated due to participant fatigue or sickness.

Data analysis

All analyses on the measured absolute and differential threshold estimates were performed in logarithmic units, but for convenience the averages are reported in m/s^2 (see Figure 18). The formulas used for the conversions are:

$$abs\_th = e^{\alpha} \tag{7}$$

$$diff\_th = e^{(\alpha + \frac{\beta}{2})} - e^{(\alpha - \frac{\beta}{2})} \tag{8}$$

where abs_th and diff_th indicates the absolute and differential thresholds in linear space respectively and α and β are the mean and the standard deviation of the cumulative Gaussian respectively, in logarithmical units.

To compare human sensitivity for upward versus downward motion at different pedestal intensities, data analysis was performed on the measured thresholds in logarithmic units. A repeated-measures ANOVA was used to test for significant differences between the 2 levels of the factor "motion direction" (upward and downward motion) and between the 4 levels of the factor "motion amplitudes", corresponding to the different pedestal motion amplitudes between 0.3 m/s^2 and 2 m/s^2. A paired samples t-test was used to compare absolute thresholds for upward and downward motions. Linear and logarithmic threshold estimates and corresponding goodness of fit were compared with 2 repeated-measures ANOVA with 1 factor ("fit type"). Effects are considered to be significant if their P value is < 0.05.

To establish the perceptual law governing human sensitivity to vertical self-motion, upward and downward data were fitted with three different models. The first model expresses Weber's Law in its general form $\Delta\Phi = k\Phi$, where Φ represents the stimulus intensity. A second model, suggested as an improvement to Weber's Law, has the form $\Delta\Phi = k(\Phi+a)$, where a represents the amount of noise that exists when the stimulus is zero (Gescheider 1997). The third model is a power law ($\Delta\Phi = k*\Phi^p$). The ability of these models to describe dependencies of upward and downward differential thresholds on motion intensity was quantified by the small-sample corrected version of the Akaike Information Criteria (AIC$_c$) (Burnham and Anderson 2004), according to the formula:

$$AIC_c = N \ln\left(\frac{\sum_i^N (y_i - f_i)^2}{N}\right) + 2k + \frac{2k(k+1)}{N-k-1} \qquad (9)$$

where N is number of data points, k is the number of model's parameters, y_i represent the i-th data point and f_i is the model prediction for the stimulus associated to y_i. AIC$_c$ provides a relative measure of the models' quality, assigning smaller values to models with a better trade-off between accuracy (prediction errors on the dataset) and complexity (high number of parameters).

3.4 RESULTS

A typical run required 133 trials on average and lasted approximately 40 minutes. In only one case did the algorithm fail to converge before 200 trials, otherwise the maximum number of trials required for one condition was 187.

Individual psychometric functions were fit in logarithmic space for each tested condition. The choice of fitting in the logarithmic space (Soyka et al. 2011; Soyka et al. 2012) or in the linear space (MacNeilage et al. 2010a; Roditi and Crane 2012) is arbitrary and, to the best of our knowledge, has never been systematically addressed. Data fitted in both domains are reported in Table 3. No significant differences were found between the goodness of fits ($F_{(1,91)}=1.27$, $p=0.26$) nor between threshold estimates ($F_{(1,91)}=0$, $p=0.95$). We chose to analyse the data obtained with the logarithmic fit for two reasons. First, the adaptive procedure relied on an on-line logarithmic fit of the psychometric function to select the most informative stimuli. Additionally, the concept of fitting in the logarithmic space is consistent with decreased sensitivity for increasing motion intensities reported here and elsewhere (Zaichik et al. 1999; Mallery et al. 2010; Naseri and Grant 2012).

Figure 18 shows that differential thresholds for vertical motion depend on pedestal amplitude ($F_{(3,27)}=54.2$, $p<0.001$) as well as on movement direction ($F_{(1,9)}=6.43$, $p=0.009$). Asymmetries between upward and downward sensitivity seem to increase with stimulus intensity, although the interaction between motion direction and amplitude is not significant ($F_{(3,27)}=1.55$, $p=0.22$).

The absolute detection thresholds obtained with a pedestal acceleration of 0 m/s^2 were 0.065 m/s^2 for upward movements and 0.067 m/s^2 for downward movements and not significantly different ($t(5)=0.033$, $p=0.975$). This last finding is consistent with previous works on detection thresholds (Melvill Jones and Young 1978) and direction discrimination thresholds (Roditi and Crane 2012).

	Pedestal (m/s^2)	Threshold (m/s^2)		Entropy (bit)	
		Lin	Log	Lin	Log
up	0	0.066 +/- 0.023	0.066 +/- 0.023	2.963 +/- 0.041	2.980 +/- 0.195
	0.3	0.118 +/- 0.072	0.096 +/- 0.031	2.413 +/- 0.440	2.863 +/- 0.246
	1.1	0.209 +/- 0.107	0.206 +/- 0.112	2.702 +/- 0.253	2.628 +/- 0.275
	1.6	0.231 +/- 0.106	0.249 +/- 0.149	2.764 +/- 0.217	2.520 +/- 0.325
	2	0.286 +/- 0.103	0.297 +/- 0.130	2.875 +/- 0.207	2.503 +/- 0.298
down	0	0.069 +/- 0.035	0.068 +/- 0.033	3.039 +/- 0.098	3.036 +/- 0.247
	0.3	0.095 +/- 0.063	0.082 +/- 0.040	2.224 +/- 0.466	2.764 +/- 0.178
	1.1	0.216 +/- 0.101	0.216 +/- 0.113	2.720 +/- 0.288	2.644 +/- 0.311
	1.6	0.198 +/- 0.096	0.211 +/- 0.138	2.658 +/- 0.163	2.423 +/- 0.326
	2	0.210 +/- 0.111	0.209 +/- 0.111	2.704 +/- 0.270	2.242 +/- 0.311

Table 3 Absolute and differential thresholds and corresponding goodness of fit (mean +/- standard deviation) as obtained by fitting psychometric functions in the linear ("Lin" column) and logarithmic ("Log" column) domain. The latters are presented transformed according to equations 7 and 8 to allow for comparison.

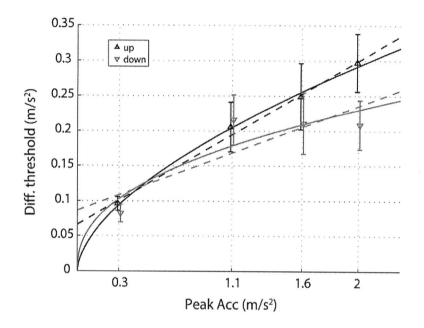

Figure 18 *Differential thresholds for upward movements (black triangles) are significantly higher than for downward movements (grey triangles). Error bars represent ±1 SEM. Their relationship with the motion intensity is more consistent with a power function (black and grey continuous lines) as opposed to a linear fit (black and grey dashed lines).*

The Weber's Laws that best fitted the collected data had coefficients k_{up}=0.16 and k_{down}=0.13 ($AIC_{c\text{-}up}$=-172.41, $AIC_{c\text{-}down}$=-172.25) For the improved Weber's Law models ($\Delta\Phi = k(\Phi+a)$) we obtained k_{up}=0.12, a_{up}=0.56, k_{down}=0.07 and a_{down}=1.16 ($AIC_{c\text{-}up}$=-173.09, AIC_{c_down}=-175.10). Finally, fitting the data with two power functions ($\Delta\Phi = k*\Phi^p$) led to the coefficients k_{up}=0.19, p_{up}=0.60, k_{down}=0.17 and p_{down}=0. 42 ($AIC_{c\text{-}up}$=-173.17, $AIC_{c\text{-}down}$=-176.39; Figure 18, solid lines). For both upward and downward directions the power law model reported the lowest AIC_c scores and is therefore the best candidate to represent the collected differential thresholds for vertical translations.

3.5 DISCUSSION

We investigated human sensitivity to vertical translations by independently measuring upward and downward differential thresholds for self-motion perception. These thresholds were found to be lower for downward compared to upward translations and to overall increase with stimulus intensity up to 0.3 m/s² at a pedestal upward acceleration of 2 m/s². According to AIC_c, this trend is best described by a power law with exponents of 0.60 and 0.42 for upward and downward motion respectively. We however point out that differences in AIC_c scores between the three compared models are often smaller than 2, suggesting only modest preference for the power laws (Burnham and Anderson 2004). The use of a power law to describe differential thresholds is attributed to Guilford (1932) and should not be confused with Stevens' power law, which rather relates magnitude estimation responses to physical stimulus intensities. Notably, despite clear changes in motion sensitivity over the wide tested range, fitting psychometric

functions in the logarithmic rather than in the linear space does not improve the quality of the fit. This suggests that, for small changes in motion amplitude, differential thresholds can be considered constant.

Upward and downward sensitivity

We find a significant difference between differential thresholds for upward and downward motion, a difference that increases with stimulus intensity. This difference is not present at the level of the vestibular afferents, where cells preferentially excited by upward translations show comparable sensitivities with those preferentially excited by downward translations (Jamali et al. 2009). It is legitimate to consider whether asymmetries in the perception of vertical motion are an immediate consequence of the nonlinear perceptual law. Since the perception of vertical movements modulates around gravity (G), the net acceleration felt by human inertial sensors during downward acceleration is smaller (G − acc) than the one felt during upward acceleration of equal intensity (G + acc). Consequently, monotonically increasing perceptual laws (such as Weber's law or the power law) predict thresholds for downward motion to always be smaller than any threshold for upward motion. However, arranging measured differential thresholds according to the net accelerations (from G − 2 m/s^2 to G + 2 m/s^2) does not allow us to fit any constant or monotonic perceptual law. Decreasing sensitivities for net accelerations diverging from gravity lead to the conclusion that the brain compensates for gravity.

Having ruled out this explanation, there are at least three other factors that could explain the asymmetry in the perception of vertical translation. First, asymmetries might derive from central processing stages and cognitive factors. For instance, a higher weight could be assigned by the brain to the perception of downward motion given its importance for detecting falls and maintaining balance. A second explanation might be related to the noise introduced by the simulator on the commanded accelerations. In fact, downward movements of the CyberMotion Simulator present a higher Signal to Noise Ratio than upward movements (Nesti et al. 2014b). This means that a downward acceleration contains less noise as opposed to a vertical acceleration of equal commanded intensity, but it is unclear whether this difference is reflected in behavioural measures. A motion analysis of the simulator employed in Jamali et al. (2009), where comparable saccular afferent sensitivity are shown for cells responding to upward and downward motion, might help clarify this point. Finally, a third explanation for the effect of motion direction on motion sensitivity may reside in a significant contribution of somatosensory cues (Seidman 2008; Seidman et al. 2009). Indeed, during surge and heave, participants usually experience asymmetric tactile stimuli according to the direction of motion, whereas sway movements act symmetrically on the human body. Asymmetries in absolute detection thresholds for heave (in head coordinates) and surge are reported from Benson et al. (1986), with footward movements and backward movements being correctly detected more frequently than headward movements and forward movements respectively. We speculate that, in our work, the use of foam padding might have reduced tactile cues during absolute thresholds measures up to the point where no significant evidence of their contribution on motion detection sensitivity could be observed. Nevertheless, the slower decrease in sensitivity that we observed for downward versus upward movement at higher motion intensities might reflect the influence of somatosensory cues

These three alternatives can only be partially resolved by testing head-vertical sensitivity in the earth horizontal plane. Asymmetries are expected to disappear if they are derived from a balance mechanism or from different simulator-dependent noise, since balance is not threatened by horizontal motion in world coordinates and the simulator's noise for horizontal motion is expected to be independent of motion direction. On the other hand, if asymmetries arise entirely or partially from asymmetric somatosensory cues they may persist for head-vertical motion in the earth horizontal plane.

The importance of vestibular relative to proprioceptive and somatosensory information for body orientation and self-motion perception has been addressed by previous studies. In animals, extremely similar balance disorder and motor incoordination were observed by Carpenter et al. (1959) in labyrinthectomized cats and by Cohen (1961) in monkeys and baboons deprived of neck proprioceptors. This indicates equally essential roles of head-to-space information (arising from the vestibular system) and of head-to-trunk information (arising from neck proprioceptors). These questions have been addressed also in human vestibular loss patients (Walsh 1961; Gianna et al. 1996; Mallery et al. 2010; Cutfield et al. 2011; Valko et al. 2012; Agrawal et al. 2013), who in theory can only rely on somatosensory and proprioceptive cues. Overall, differences between healthy participants and patients show high variability, suggesting that the amount of information coming from non-vestibular sources varies with the experimental conditions, experimental setup, and motion profiles. For instance, Valko et al. (2012) showed thresholds from 1.3 to 56.8 times higher for vestibular loss patients than for healthy participants, depending on the frequency and the type of motion. Although it seems that the vestibular system plays a primary role in perceiving self-motion, from the observation that vestibular loss patients are still able to perceive motion in darkness we can safely conclude a contribution of the somatosensory and proprioceptive cues. This might explain asymmetric vertical sensitivity for upward and downward motion if these cues, asymmetric for heave motion, are noticeably different.

Vertical self-motion sensitivity

Two power laws well describe the increase in upward and downward differential thresholds upon stimulus intensity (Figure 18). In contrast, such a nonlinear dependence on motion intensity is not observed over the tested range by the saccular afferent fibres of the vestibular system in squirrel monkeys and rhesus monkeys (Fernández and Goldberg 1976b; Jamali et al. 2009) nor in human eye movements, which maintain a constant level of accuracy and precision over a wide range of rotation intensity (Pulaski et al. 1981; Weber et al. 2008). Perceptual nonlinearities must therefore arise at a further stage along the neuronal path that process self-motion information and are perhaps due to central processing of the vestibular signals, multisensory integration processes and/or cognitive factors. The discrimination capability for high stimulus intensities is remarkable compared to other human sensory systems which are often best characterized by exponents close to 1 (Teghtsoonian 1971), i.e. the linear relationship between stimulus intensity and differential threshold described by Weber's Law. As suggested by Mallery et al. (2010), the deviation from Weber's law due to heightened sensitivity at larger stimulus intensities may be related to the role of the vestibular system for maintaining posture even at these high stimulus intensities.

Note that, for downward movements, one could almost argue that differential thresholds are independent of stimulus intensity, as only for the 0.3 m/s^2 pedestal was the threshold significantly different than those for the other pedestals (Figure 18). However, we still favour the power law over a conclusion of stimulus-independent downward sensitivity for several reasons. First, the data consistently show a clear difference at 0.3 m/s^2 and confirm the differential threshold reported in MacNeilage et al. (2010a). Second, it is more parsimonious to describe upward and downward sensitivity with the same function, which also leads to better AIC_c scores ($\text{AIC}_{c\text{-down}}$=-170,07 when fitting downward differential thresholds with their mean). Finally, dependencies are to be expected based on previous studies on self-motion sensitivity (Zaichik et al. 1999; Mallery et al. 2010; Naseri and Grant 2012).

A technical consideration is however necessary: for the motion simulator used here, as well as for other common motion simulators (AGARD 1979), the quality of the reproduced signals, expressed in terms of Signal to Noise Ratio (SNR), increase with motion intensity (Nesti et al. 2014b). Thus it is possible that some amount of change characterized here by the exponents of the perceptual laws actually reflects variation in stimulus quality. Note that this is likely to be a very general problem, not specific to the current study. Similar SNR is expected for other simulators as well. However, none of the physiological studies mentioned above (Fernández and Goldberg 1976b; Weber et al. 2008; Jamali et al. 2009) reported changes in sensitivity, suggesting that changes in the simulator SNR over the tested motion range are not picked up by the vestibular system.

Heave differential thresholds have been investigated in the past by MacNeilage et al. (2010a) and Zaichik et al. (1999). In the first study a 6 degrees of freedom motion platform was used to generate 1 Hz sinusoid-like acceleration profiles. The differential threshold for a pedestal of 0.3 m/s^2, measured with a 2IFC experimental design similar to the one we employed, is reported to be 0.117 ± 0.078 m/s^2. They also measured a direction discrimination threshold of 0.097 ± 0.034 m/s^2 for vertical motion in a single interval discrimination task, where participants were asked to correctly identify the direction of motion (upward or downward). This value has to be multiplied by $\sqrt{2}$ before comparing it with the results of a 2-interval discrimination task. Further accounting for the different threshold definition (84% rather than 75% correct answer probability) we obtain a comparable absolute threshold of 0.093 m/s^2. Given the procedural similarities between this and the present study, it is possible to quantitatively compare our measurements with MacNeilage et al. (2010a). Results agree for the differential thresholds at 0.3 m/s^2, but our detection threshold of 0.066 m/s^2 is lower than their direction discrimination threshold by approximately 30% after correcting for the different definitions of absolute threshold (84% rather than 75% correct answer probability). This discrepancy can likely be attributed to the different experimental task. Indeed, thresholds are known to be higher for motion direction discrimination rather than motion detection, especially in the vertical direction (Melvill Jones and Young 1978). In Zaichik et al. (1999) a very different methodology was employed and a comparison of the measured thresholds would be meaningless. However, the Weber's perceptual law they found for vertical movements in the range of $0 - 0.6$ m/s^2 is qualitatively consistent with our findings.

Absolute threshold can be defined as the differential threshold relative to a pedestal of zero. Unfortunately, the absolute thresholds measured in this study cannot be directly compared with differential threshold measured for non-zero pedestal values because of differences in the experimental design. However, if comparable methods had been used, we would expect the differential threshold for 0 m/s^2 pedestal to be equal to or greater than the absolute threshold reported here. This would represent a significant deviation from the power law fits illustrated in Figure 18. In fact, it would not be unreasonable to expect a decrease (or dip) in differential thresholds as pedestal increases from zero to small non-zero values, before increasing again with increasing pedestal. Such "dipper functions" are commonly observed in other psychophysical domains (Solomon 2009). The shape of the differential threshold curve from pedestal 0 m/s^2 to 0.3 m/s^2 represents an interesting topic for future research.

Results from many psychophysical studies (Benson et al. 1986; Zaichik et al. 1999; MacNeilage et al. 2010a; Roditi and Crane 2012; Valko et al. 2012; summarized in Table 4) agree that absolute thresholds are about 2 times higher for vertical motion than for horizontal motion. Absolute thresholds for horizontal linear motion have been previously investigated by Soyka et al. (2011) using the same simulator and a very similar setup. Although they did a direction discrimination rather than a detection task and at different frequencies (0.17, 0.42, 0.67 Hz), a comparison is still possible using the model they propose to account for frequency dependencies. Here, a direction discrimination threshold of about 0.02 m/s^2 is predicted for horizontal sinusoidal acceleration profiles at 1 Hz, a value lower than the detection threshold we measured for vertical motion (0.066 m/s^2), providing further evidence that humans are less sensitive to vertical motion.

By comparing our results with Naseri and Grant (2012) we show that differences in sensitivity to horizontal and vertical motions exist not only at absolute threshold level but also for supra-threshold movements. Indeed, our data show differential thresholds for vertical movements that are always higher than those measured by Naseri and Grant (2012) for horizontal movements over the same motion range (up to 2 m/s^2). The lower frequency used for their stimuli (0.25 to 0.6 Hz) supports this conclusion even more, since self-motion perceptual thresholds are known

	Task	Stimuli	Threshold (m/s²)		
			Surge	Sway	Heave
Benson et al. 1986	Discrimination	0.33 Hz sinusoidal acc	0.06	0.06	0.15 *
Zaichik et al. 1999	Detection	0.95 Hz sinusoidal acc	0.03	0.05	0.08
MacNeilage et al. 2010	Discrimination	1 Hz sinusoidal acc	-	0.06	0.10
Roditi and Crane 2012	Discrimination	1 Hz sinusoidal acc	0.03	0.03	0.08
Valko et al. 2012	Discrimination	1 Hz sinusoidal acc	-	0.02	0.05
Soyka et al. 2011 (model)	Discrimination	1 Hz sinusoidal acc	0.02 **	-	-
Present data	Detection	1 Hz sinusoidal acc	-	-	0.07

* participants were lying on their back, which MacNeilage et al. (2010a) have shown raises perceptual thresholds

** data were not significantly different for surge and sway, and were therefore pooled before fitting the model

Table 4 *Absolute thresholds for 3D translations. Differences between measures are due to the task and the different simulators employed (Nesti et al. 2014b), but within studies thresholds are consistently higher for vertical movements. Model prediction based on Soyka et al. (2011) and our results are grouped since the same simulator was employed.*

to decrease at higher frequencies (Benson et al. 1986; Soyka et al. 2011). It has been suggested that lower thresholds for horizontal than for vertical linear motion might derive from differences in the otolith organs response to linear acceleration acting on different axis of the head (Benson et al. 1986). Neurophysiologic measures of the primary afferent neurons in the squirrel monkey show sensitivity (spikes/s/g) about 30% higher for horizontal than for vertical movements (Fernández and Goldberg 1976c). More recent neuronal findings (Jamali et al. 2009; Yu et al. 2012) report however similar gains for otolith afferents of the rhesus monkey when responding to horizontal and vertical translation. It is therefore not clear if higher perceptual thresholds for vertical as compared to horizontal translations partially reflect a property of the vestibular afferents, but the perceptual discrepancies are anyway higher than what the neurophysiologic data from Fernández and Goldberg (1976c) would predict. This incongruence between objective and subjective data could again reside in higher level processing of the vestibular signals, in multisensory integration or in cognitive factors. As was suggested above, a possible interpretation for vertical asymmetries is that the central nervous system might favour horizontal sensitivity more than vertical sensitivity since it is more informative for balance control.

ACKNOWLEDGMENTS

We gratefully thank Karl Beykirch, Michael Kerger and Harald Teufel for technical assistance and scientific discussion.

GRANTS

AN and MB-C were supported by funds from the Max Planck Society. PRM was supported by the German Federal Ministry of Education and Research under the Grant code 01 EO 0901. This work was also supported by the Brain Korea 21 PLUS Program through the National Research Foundation of Korea funded by the Ministry of Education. The funders had no role in study design, data collection and analysis, decision to publish, or preparation of the manuscript.

COPYRIGHT

4

SELF-MOTION SENSITIVITY TO VISUAL YAW ROTATIONS IN HUMANS

This chapter has been reproduced from an article published in Experimental Brain Research: Nesti A, Beykirch KA, Pretto P, Bülthoff HH (2015) Self-motion sensitivity to visual yaw rotations in humans. Exp Brain Res vol 233(3), pp 861-899.

4.1 ABSTRACT

While moving through the environment, humans use vision to discriminate different self-motion intensities and to control their actions (e.g. maintaining balance or controlling a vehicle). How the intensity of *visual* stimuli affects self-motion perception is an open, yet important, question. In this study, we investigate the human ability to discriminate perceived velocities of visually induced illusory self-motion (vection) around the vertical (yaw) axis. Stimuli, generated using a projection screen (70x90 degrees Field of View), consist of a natural virtual environment (360 degree panoramic colour picture of a forest) rotating at constant velocity. Participants control stimulus duration to allow for a complete vection illusion to occur in every single trial. In a two-interval forced-choice task, participants discriminate a reference motion from a comparison motion, adjusted after every presentation, by indicating which rotation feels stronger. Motion sensitivity is measured as the smallest perceivable change in stimulus intensity (differential threshold) for 8 participants at 5 rotation velocities (5, 15, 30, 45 and 60 deg/s). Differential thresholds for circular vection increase with stimulus velocity, following a trend well described by a power law with an exponent of 0.64. The time necessary for complete vection to arise is slightly but significantly longer for the first stimulus presentation (average 11.56 s) than for the second (9.13 s), and does not depend on stimulus velocity. Results suggest that lower differential thresholds (higher sensitivity) are associated with smaller rotations, because they occur more frequently during everyday experience. Moreover, results also suggest that vection is facilitated by a recent exposure, possibly related to visual motion after-effect.

4.2 INTRODUCTION

When moving through the environment, continuous variations of the retinal image (optic flow) often provide important self-motion cues and play a major role in self-motion perception (von Helmholtz 1925; Gibson 1950). The importance of optic flow is particularly striking when conflicting information arises from the sensory systems involved in the perception of self-motion (mainly visual, vestibular and somatosensory systems). A frequently cited experience is the feeling of self-motion on a stationary train when a neighbouring train begins to move.

In everyday life, the intensity of self-motion varies over a wide range, from small subtle postural changes to stronger movements occurring during sport activities, for instance. Reliable estimates of these movements are obviously essential for a variety of crucial tasks (e.g. maintaining posture in presence of external disturbances or controlling a vehicle). However, recent studies showed that the ability to estimate motion intensity varies with the intensity of motion stimuli (Zaichik et al. 1999; Mallery et al. 2010; Naseri and Grant 2012; Nesti et al. 2014a). In these studies differential thresholds (DTs), i.e. the smallest detectable changes in stimulus intensity, were measured over wide ranges of linear (e.g. 0-2 m/s^2) and angular (e.g. 0-160 deg/s) motions. Results indicate that the relationship between DTs and stimulus intensity may be described by a power law, $\Delta S = k * S^{b}$, where S is the stimulus intensity, ΔS is the DT, and k and b depend on the type of investigated motion (e.g. vertical translation or rotation). These works unequivocally show that DTs for self-motion in darkness are *not constant* but rather increase with motion intensity. In other words, motion sensitivity (i.e. the ability to detect small changes in stimulus intensity) worsens at higher motion intensities. This implies a *nonlinear* relationship between actual motion intensity and perceived motion intensity (Fechner 1860). Indeed, any sensory system whose sensitivity is *not constant* over the response range of the sensor is by definition *nonlinear*. This perceptual nonlinearity, shown for other sensory modalities (see e.g. Teghtsoonian (1971)), might reflect better sensitivity for ranges of stimulus intensities that are more frequent in everyday life (Stocker and Simoncelli 2006).

Despite the well-established role of visual cues in self-motion perception (Dichgans and Brandt 1973), less effort has been dedicated to measuring DTs for visual self-motion cues, perhaps because of the methodological challenge of ensuring that visual motion is indeed perceived as self-motion rather than object motion. DTs for visually simulated motion in depth supporting a Weber-like perceptual law were measured by Wei et al. (2010), to investigate how visual perceptual uncertainty affects balance responses. However, their choice for visual stimulation may indeed not fully address the issue of self-motion versus object motion for two reasons. First, the radial velocity of the random-dot flow field did not vary with eccentricity, as it would for vision during self-motion in a 3D environment. Second, the stimulus duration of 0.8 sec is too short for a compelling visually-evoked self-motion illusion (vection), which usually requires between 2 and 40 seconds, depending on the experimental conditions (see Discussion). In this study, we investigate human self-motion sensitivity by measuring DTs for visually-evoked yaw rotation perception in an immersive virtual environment (circular vection). This constitutes a step forward in the understanding of self-motion sensitivity in more realistic conditions, where inertial and visual cues are both available. We hypothesize that DTs for vection increase with motion intensity following a trend described well by a power law, confirming the nonlinearity of self-motion perception observed when moving in darkness. This experiment will furthermore pave the way for a comparison between DTs for different combinations of visual and inertial cues and will lead to further investigation of the neural processes underlying self-motion perception and multisensory integration.

Studying the contribution of visual cues to self-motion perception requires great care in the design of the stimulus. Indeed, as previously mentioned, moving visual stimuli do not necessarily evoke a perception of self-motion. A well-established theory argues for a reciprocal relationship between the perception of object motion and self-motion (Dichgans and Brandt 1978), although other models have been suggested (Wertheim 1994). Moreover, several studies show that vection is modulated by the physical properties of the stimulus, as well as by cognitive factors. For example, the vection onset time (VOT) depends on the field of view of the visual stimulation and on the numbers of elements (e.g. dots) in the scene (Webb and Griffin 2003) while the sensation of vection may be enhanced by adding inertial vibration (Riecke et al. 2005) or by using a realistic, as opposed to unnatural, virtual environment (Riecke et al. 2005; Riecke et al. 2006). In this study, a realistic visual stimulus rotating at constant velocity around the vertical axis of the participants was used. Note that constant velocity rotations are a particularly appropriate choice, as the insensitivity of the vestibular system to constant rotations strongly mitigates conflicting multisensory information. Moreover, participants also experienced stimulus-unrelated vibrations throughout the entire experiment (see Methods and Discussion).

Psychophysically measuring human self-motion DTs will allow improvements in the field of self-motion perception modelling, where the lack of experimental evidence has often led to the assumption of constant motion sensitivity (see e.g. Bos and Bles (2002), Zupan et al. (2002), Newman et al. (2012)). Improved model accuracy is beneficial for the design of simulation environments, such as motion simulators used for driver and pilot training. For instance, in vehicle simulation the motion of the simulated vehicles could be modified to better suite simulator capabilities as long as the manipulations remain unperceived (i.e. below DT). A better understanding of pilots' perception over wide motion intensities also allows for more effective simulator training protocols in extreme conditions with the goal of better prediction and avoidance of accidents. Furthermore, in the medical diagnosis of balance disorders, psychophysical tests (Merfeld et al. 2010) could supplement currently used eye movement tests (Bárány 1921; Halmagyi and Curthoys 1988) in cases where eye movement cannot be measured (Merfeld et al. 2010) or to specifically measure perception since self-motion perception and ocular reflexes do not always match (MacGrath et al. 1995; Merfeld et al. 2005a; Merfeld et al. 2005b; Wood et al. 2007).

4.3 METHODS

Participants

Eight participants (aged 26-53 years, 1 female), five naïve and three authors (AN, KB and PP) took part in this study and gave their informed written consent prior to inclusion in the study, in accordance with the ethical standards specified by the 1964 Declaration of Helsinki. They all had normal or corrected-to-normal vision and reported no history of balance disorders and no susceptibility to motion sickness.

Setup

The study was conducted using the Max Planck Institute CyberMotion Simulator (Figure 19, for technical details refer to Nieuwenhuizen and

Bülthoff (2013); KUKA Roboter GmbH, Germany). Inside the closed cabin, two projectors (1920x1200 pixels resolution, 60 Hz frame rate) display on the white, curved inner surface of the cabin door, approximately 60 cm from the participants' head. A field of view (FOV) of approximately 70x90 deg and an actual resolution of approximately 19.6 pixels/deg were used for the experiment. Participants were seated in a chair with a 5-point harness. They wore headphones playing white noise during the presentation of visual stimuli (Figure 19 and 21) to mask external auditory cues. The head was restrained with a Velcro band, which combined with careful instruction to maintain an upright posture, helped participants to avoid coriolis effects (Guedry and Benson 1976; Lackner and Graybiel 1984), i.e. the sense of discomfort or nausea following head tilts during inertial rotations at constant velocity (see below). Bi-directional participant-experimenter communication was active throughout the experiment for safety. Participants interacted with the experiment using a button box with 3 active buttons, one to initiate and terminate the stimulus (control button) and the other two for providing a forced choice response (response buttons, see procedure). Eye movements were not recorded.

Stimulus Generation

Visual stimuli were generated with authoring software for interactive 3D applications (Virtools, 3DVIA). A 360 degrees panoramic picture of a forest (Figure 20) was displayed on the surface of the cabin (60 cm away). In order to avoid motion parallax the software projected the image on a cylinder created in the virtual environment whose axis coincides with the earth-vertical axis of the participants' point of view in the virtual environment. The radius of the virtual cylinder (5 meters) was chosen to achieve a satisfactory visual appearance on the screen (i.e. texture resolution and object size).To preserve the participants' natural behaviour, no visual fixation was used. Stimuli consisted of rotations of the virtual cylinder around its axis with constant rotational velocities in the range of 5-72

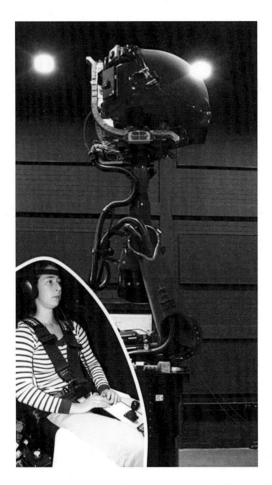

Figure 19 *Experimental setup. Participants sat inside the simulator's cabin, where visual stimuli were presented on the inner surface of the cabin door by means of 2 projectors (stimulus resolution: 19.6 px/deg, refresh rate: 60 Hz, FOV: 70x90 deg).*

Figure 20 *Fragment of the 360 degree panoramic colour picture used for generating the realistic virtual environment.*

deg/s. A constant linear acceleration onset/offset was generated, lasting two seconds for the reference stimulus and slightly longer for the comparison due to its higher velocity (see Figure 21). The onset/offset ramp resulted in a more tolerable and natural motion sensation, as compared to a step onset/offset.

During each session, participants were continuously rotating around the head-centred vertical axis at the constant velocity of 20 deg/s so to generate vibrations unrelated to the stimulus. Note that the perception of constant inertial rotations disappears within a few seconds after rotation onset (Bertolini et al. 2011), and even the small otoliths stimulation is sub-threshold[5]. However, the vibrations resulting from the simulator's motion have been shown to enhance VOT and convincingness (Riecke et al. 2005). The absence of centripetal accelerations was monitored with a 3-axis accelerometer placed on the top of a participant's head. Rotation direction was reversed approximately every 15 min corresponding to session breaks (see below), and stimulus presentation began 1 minute after constant velocity was reached, allowing for the sensation of rotational motion to disappear. To avoid confusion, throughout the chapter we refer to the visual stimuli in a frame of reference relative to the rotating participants.

Procedure

Each trial was composed of two consecutive presentations of the visual stimulus, separated by a pause of 3 seconds (see Figure 21). The constant velocity amplitude of one of the presentations (reference stimulus) remained the same across all trials while the amplitude of the comparison was systematically varied. Reference and comparison had opposite directions, as this was found to hinder comparison of purely visual (object)

---

[5] Assuming that the otolithic organs are located at 10 cm from the center of rotation, the centripetal acceleration sensed by the otoliths is: $F = r * (\text{angular velocity})^2 = 0.1$ m $* (20 \text{ deg/s} * \text{pi} / 180 \text{ deg})^2 = 0.012$ m/s^2. This constant acceleration is below perceptual thresholds (see (Nesti et al. 2014a) for a review).

velocities in pilot work. Such a comparison might in fact artificially lower the thresholds, facilitating the discrimination task without however contributing to self-motion perception. Presentation order was randomized to prevent complications due to order effects and visual motion after-effects (Hershenson 1989).

Prior to each trial, the virtual environment was visible and stationary in front of the participants. They initiated each trial by pressing the control button. The visual environment was then rotated at constant velocity. Participants were in control of the stimulus duration and were instructed to terminate it by pressing the control button when they confidently perceived the virtual scene as stationary. According to the "reciprocity" theory between object and self-motion (Dichgans and Brandt 1978), this is equivalent to confidently perceiving themselves being rotated within a *stationary* scene, with all the visual motion attributed to self-motion. After both stimuli of the trial were terminated, the screen turned black and participants were asked to report "which rotation was faster" by pressing one of the two response buttons (first or second). No feedback on the correctness of the response was provided. The time to scene stationarity (TSS), here defined as the time between stimulus onset and stimulus termination, was recorded for each stimulus with a resolution of 1 ms. After each trial, participants were allowed to rest, the virtual environment remained visible and stationary in front of them and no white noise was presented.

The experiment was divided into 5 sessions of approximately 45 minutes each, with breaks of approximately 5 minutes every 15 minutes of experiment to avoid fatigue. Participants completed the experiment over 5 different days (1 condition per session per day, order randomized). In each condition, the participants' DT was measured for a different reference velocity (5, 15, 30, 45 and 60 deg/s) using a psychophysical two-interval forced-choice (2IFC) procedure. Comparison velocities were adjusted for every trial according to an adaptive staircase algorithm which decreased the stimulus level after 3 consecutive correct responses and increased it after every incorrect response (3-down 1-up rule (Levitt 1971)). This algorithm eventually converges to the stimulus level where a stimulus increase (wrong answer) or decrease (3 consecutive correct answers) are equally probable (p=0.5), meaning that the probability of a single correct answer is 0.794 (cubic root of 0.5). For the 5 deg/s reference condition the initial comparison velocity was 7 deg/s, with a step size of 0.2 deg/s, halved after the staircase reversed direction 5 times. For the other conditions the initial comparison velocity c_0 was set according to the formula $c_0 = 6 / 5 *$ ref_v, where ref_v is the reference velocity. The step size, initially set at 2 deg/s, was halved after 5 reversals (1 deg/s) and again after 10 (0.5 deg/s). Every session was terminated after 13 reversals. The comparison velocities over one experimental session for a typical participant are illustrated in Figure 22.

Data analysis

DTs were obtained from each condition by averaging the last 8 staircase reversals (Figure 23). An alternative estimate of DTs was obtained from the Least-Squares Estimation (LSE) of a psychometric function to the participant's responses for every condition. In this case, DTs were defined as the reference velocity increment necessary for a 0.794 probability of correct discrimination. This allowed for investigating, using an ANOVA, whether DT estimates could be affected by the formula chosen for their calculation. Linear regression analysis was performed on DTs to test whether human vection sensitivity depends on motion intensities. A repeated-measures ANOVA was used to test for differences in the TSSs between the 2 levels of the factor "presentation order" (first or second) and between the 5 levels of the factor "stimulus intensity", corresponding to the 5 different reference velocities. Statistical analyses were performed in MATLAB (2012a) using the Statistical Toolbox. Effects are considered statistically significant if their P value is < 0.05.

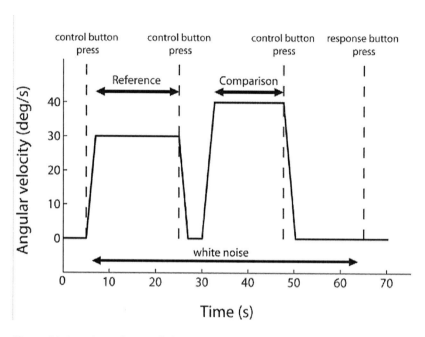

Figure 21 *Angular velocity of the virtual stimulus during a typical trial. Note that the order was varied randomly (see text).*

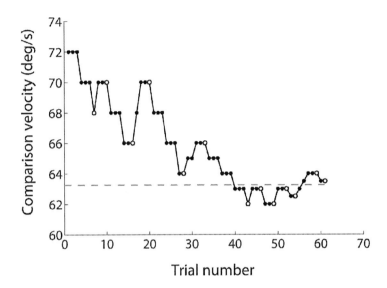

Figure 22 *Comparison velocities for each trial of a typical experimental session with a reference velocity of 60 deg/s. Empty circles indicate reversal points. The last eight reversals were averaged to compute the DT (dashed line).*

Two different models, a Weber's law (Fechner 1860) and a power law (Guilford 1932), proposed in the literature to relate DTs to stimulus intensity, were fit to the data. The Weber's law function has the general form $\Delta S = k * (S + a)$, where ΔS is the DT, S the stimulus intensity, k the Weber fraction and a represents the amount of noise that exists when the stimulus is zero (Gescheider 1997), while the power law function has the form $\Delta S = k * S^b$. The two models describing rotational vection sensitivity were compared based on their goodness of fit, measured by the coefficient of determination r^2.

4.4 RESULTS

Each condition took approximately 45 minutes and required on average 63 trials. No session needed to be terminated because of fatigue or other reasons, although mild symptoms of motion sickness were often reported (see Discussion).

As illustrated in Figure 23, no difference is found between DTs obtained by reversal averaging and by LSE ($F(1,7)=2.62$, $p=0.15$). We therefore proceeded to analyse the former estimates, as they don't require an assumption on the shape of the psychometric function. Indeed, such an assumption cannot be properly done because the adaptive procedure concentrates stimulus presentations only around its region of convergence (0.79 probability of correct discrimination, see Methods).

DTs for vection are presented in Figure 24. Regression analysis yielded a slope coefficient of 0.11 ± 0.017, indicating that DTs increase with motion intensity ($t(38)=6.30$, $p<0.001$).

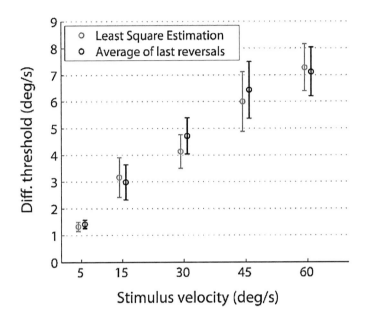

Figure 23 *Comparison between DTs estimated by LSE and by the average of the last 8 reversals. Error bars represent ±1 SEM.*

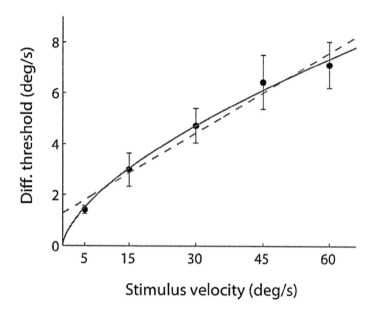

Figure 24 *DTs for head-centred yaw rotations increase with stimulus intensity following a trend well described by a concave power law (continuous line). A Weber's Law function (dashed line) provides a slightly poorer fit. Error bars represent ±1 SEM.*

Fitting the data with a Weber's law function in the form $\Delta S = k * (S + a)$ resulted in the coefficients $k = 0.11$ and $a = 12.12$, whereas from the fit of a power law function ($\Delta S = k * S^b$), the coefficients $k = 0.54$ and $b = 0.64$ were obtained. Coefficients of determination are $r^2 = 0.97$ and $r^2 = 0.99$ for the Weber's and power law respectively, indicating that the both functions provide a good description of the measured DTs.

Average TSSs (Figure 25) do not significantly depend on the motion intensity ($F(4,28)=0.16$, $p=0.96$), however they significantly depend on presentation order ($F(1,7)=10.43$, $p=0.015$), with shorter TSSs for the second stimulus of each trial (see Fig. 7). Average TSSs across all conditions are 11.56 and 9.13 seconds for the first and second stimulus presented, respectively.

4.5 DISCUSSION

In this study, we investigated human sensitivity to visually induced self-motion perception as evoked by immersive visual stimulation. We found that DTs increase with stimulus intensity (i.e. rotational velocities), indicating that sensitivity to circular vection is not constant over the investigated motion intensity range, but rather worsen at greater velocities. This represents a nonlinearity in the perception of self-motion. Such perceptual nonlinearity also emerges from psychophysical studies on linear vection discrimination (Wei et al. 2010) and magnitude estimation (Brandt et al. 1973; Dichgans and Brandt 1973). Perceptual nonlinearities may also explain why postural responses evoked by visual stimulation do not continuously increase with stimulus amplitude but rather saturate (van der Kooij et al. 2001; Wei et al. 2010). As suggested by Wei et al. (2010), this could reflect a Bayesian integration by the CNS of visual and inertial cues, assigning less weight to stronger as compared to weaker visual motions as they have greater uncertainty (i.e. higher DTs). Interestingly,

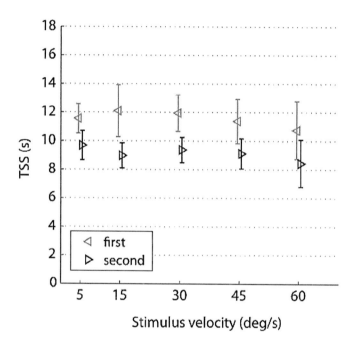

Figure 25 *TSSs are on average independent from the visual motion intensity. Participants took a statistically significantly longer time for terminating the first rather than the second stimulus. Error bars represent ±1 SEM.*

nonlinear behaviour in response to rotational optic flow is not shown in electrophysiological recordings from neurons in MSTd (Tanaka and Saito 1989) and in the vestibular nuclei (Dichgans et al. 1973; Henn et al. 1974; Waespe and Henn 1977), where average firing rates linearly depend on stimulus velocity. Similarly, vestibular and optokinetic reflexes, responsible for stabilizing the gaze in response to head rotations, show an approximately linear response over the investigated range of visual (Paige 1994) and inertial (Pulaski et al. 1981; Weber et al. 2008) rotational velocities. A possible reconciliation between physiological linearity and perceptual nonlinearity might relate to the increased firing rate variability observed in the vestibular nuclei for stronger compared to weaker self-motion intensities (Dichgans et al. 1973; Henn et al. 1974; Allum et al. 1976; Waespe and Henn 1977). Indeed, higher variability of the sensory signals leads to less precise perceptual discriminations (higher DTs) even when the average sensory response (firing rates) over multiple repetitions remains linear. Further investigation is however required in order to relate neurophysiological responses to psychophysical thresholds for self-motion.

Both the power law model and the Weber's law model provided an excellent fit for average DTs over the investigated motion range. Nevertheless, it is known that for a large range of sensory input amplitudes, Weber's law does not hold (Teghtsoonian 1971) and the power law captures the changes in DTs better (Guilford 1932). This suggests that differences in model accuracy will arise when measuring DTs for stronger and/or weaker yaw rotation intensities, with the power law model becoming the preferable alternative. We did not measure DTs for reference velocities higher than 72 deg/s (corresponding to 1.2 deg/frame in our setup) because, as confirmed by preliminary testing on two different participants, visual rotations faster than 1.2 deg/frame resulted in a visual blurriness that prevents vection from arising in our setup. Furthermore, we noticed that rotations slower than 5 deg/s did not succeed in evoking full vection within reasonable stimulus exposure times (approximately 60 seconds). Note however that both models allow one to safely conclude that

DTs significantly increase with stimulus intensity and thus that the perception of self-motion from visual cues is nonlinear.

Self-motion perception arises from central processing of inertial (vestibular and somatosensory) and visual information. This also includes dissociation of the visual features related to self-motion (optic flow) from those related to other events (e.g. object motion). Therefore, in a virtual environment, a coherent perception of self-motion is only possible through a careful design of the structural and temporal properties of the visual stimulus, where the design should also consider their interaction with stimuli of other sensory modalities. In this study, particular care was taken to address this point, resulting in the following choices. The use of a forest rather than vertical bars (Brandt et al. 1973) or a dot field (Berthoz et al. 1975) was motivated by studies showing that a natural stimulus decreases VOT and TSS, and increases immersion in a vection task (Riecke et al. 2005; Riecke et al. 2006). A ramp onset of the visual motion was chosen over a step onset because, according to outlier detection models of perception (Wei and Körding 2009), sudden changes of the visual environment (step onset) are more likely to be due to movements of the surrounding and therefore neglected for estimating self-motion, whereas gradual changes in self-motion velocities are a more natural self-motion stimulus. Subjective reports of two participants experiencing both types of onset additionally confirmed that a ramp onset generates a more tolerable multisensory conflict. We employed constant velocity visual rotations around the vertical axis of the participants, a stimulus that minimizes sensory conflicts (see below) and therefore favours a coherent self-motion perception. The lack of visual fixation, combined with careful instruction to look ahead, favoured immersion in the virtual environment, while at the same time allowed for peripheral stimulation and optokinetic nystagmus, thereby avoiding complications due to the Aubert-Fleischl paradox (de Graaf et al. 1991). Given the considerable individual differences in VOTs (Brandt et al. 1973; Berthoz et al. 1975; Riecke et al. 2005; Riecke et al. 2006), the duration of stimulus presentations was self-paced, i.e. participants were instructed to terminate the stimulus only after perceiving the visual scene as stationary (complete self-motion illusion). Finally, stimulus offset also followed a ramp profile, with the visual stimulus slowing down to a stationary visual pattern. This, again, resembles a more natural self-motion pattern than a sudden stop and mitigates both nausea and visual motion after-effect. Overall, we believe that this paradigm is well-suited for evoking vection and sets a useful precedent for self-motion studies in visual environments.

Average TSSs were 11.56 and 9.13 seconds for the first and second stimulus presented, respectively. These values are consistent with Brandt et al. (1973), where a pure sense of self-rotation was perceived on average 8-12 seconds after stimulus onset. Note however that average VOTs in the literature show high variability, ranging between 2 and 40 seconds depending on stimulus properties and cognitive factors such as visual FOV, vibrational cues and scene naturalism (Brandt et al. 1973; Berthoz et al. 1975; Webb and Griffin 2003; Riecke et al. 2005; Riecke et al. 2006). Similar dependences are therefore to be expected also in TSSs. In agreement with VOTs reported by Brandt et al. (1973) and Berthoz et al. (1975), we found no significant dependency of TSSs on stimulus intensity. This result differs, however, from the result found by Riecke et al. (2006). We additionally report a significant effect of presentation order on TSSs, suggesting that vection arises more easily shortly after previous exposure. A similar observation is reported in Berthoz et al. (1975). The visual motion after-effect is a reasonable explanation for this result: any potential residual motion perception following the first stimulus is expected to shorten the TSSs. In this study, the inter-trial pause of 3 seconds was chosen as a compromise between keeping a vivid impression of the 1st stimulus and the 4 second time constant of the visual motion after-effect reported by Hershenson (1989) for a 20 second constant rotation stimulus. It is therefore possible that for some trials the residual self-motion perception after the first stimulus had not yet decayed to 0 deg/s when the second stimulus began. Note however that this has no repercussion on DT estimates as it does not affect the perceived self-motion intensity at the moment of the button press (i.e. when the illusion is complete and all the visual motion is attributed to self-motion).

During the course of each experimental session, all participants reported mild symptoms of motion sickness. They consistently reported the discomfort being provoked by the onset/offset of the visual stimuli, whereas the constant velocity of the visual rotations was better tolerated. This fact is explained by vestibular and somatosensory dynamics (which respond to inertial accelerations) and the sensory conflict hypothesis (Beadnell 1924) underlying the most widely accepted motion sickness theories and models (Reason 1969; Oman 1982). According to this theory, motion sickness arises whenever visual and vestibular sensory cues deviate from normal daily patterns and conflict with each other. Such a conflict was minimized to a great degree in our study by employing visual and inertial rotations at constant velocity, where a lack of angular and linear inertial accelerations (other than gravity) allows for non-conflicting sensory information from the visual and inertial sensory systems. However, the onset/offset of the visual stimulus presented a visual acceleration not matched by any physical acceleration, thus resulting in sensory information interpreted as conflicting by the central nervous system.

In this study we chose to measure DTs for stimuli rotating at constant velocity (0 Hz) as higher frequencies of the visual stimulus would likely generate a sensory conflict which prevents participants from perceiving self-motion. However, constant rotations do not frequently occur in everyday life and it is therefore legitimate to question the generalizability of the results. Constant visual rotations elicit sustained neural responses in the vestibular nuclei of alert monkeys, whereas responses to transient visual velocities are attenuated (Waespe and Henn 1977). This behaviour is often referred to as low-pass behaviour because only the lower frequencies of the input signal are maintained in the response. As demonstrated by Robinson (1977) using a modelling approach, neural responses in the vestibular nuclei to visual-only and inertial-only rotations (Waespe and

Henn 1977) add linearly, with responses to rotations in darkness showing a high-pass behaviour complementing the low-pass behaviour of the visual responses. Consequently, rotations in the light show both transient and sustained activity. This would indicate that vision mainly contributes to the perception of self-motion at low frequencies (beginning to attenuate with increasing frequencies at about 0.03 Hz (Robinson 1977)), where the response of the inertial systems (e.g. the semicircular canals of the vestibular system) is either strongly attenuated or absent (Robinson 1977). Results from the present work are therefore expected to generalize well to transient profiles in a range of frequencies present in natural movements (Grossman et al. 1988). It should be noted that visual stimuli might contain additional information that, although not directly related to the perception of self-motion, could help in discriminating different motion intensities. For instance, short or periodic movements allow to judge the intensity of the motion based on the travelled distance of specific image features (e.g. a tree), a visual task that is not informative about the self-motion experienced by the participants, who could be able to perform such task even without perceiving any self-motion. We expect that, in presence of such cues, DTs will be artificially lower; this does however not detract from the main conclusion that self-motion perception nonlinearly depends on motion intensity.

Nonlinearities in the perception of inertial, rather than visual, yaw rotation stimuli have also been investigated (Mallery et al. 2010). The measured DTs, defined as we did here at 79% chances of correct discrimination, are only slightly smaller (approximately 2, 3 and 5 deg/s for reference stimuli of 20, 40 and 60 deg/s respectively) and are similarly described with a convex power law. These similarities suggest a common neural mechanism acting on the internal representation of self-motion within the central nervous system. Future studies need to systematically compare self-motion sensitivity for inertial-only, visual-only and congruent visual-inertial cues. Beside the high ecological validity of these studies (natural movements often provide multisensory cues over wider intensity ranges), such comparison will inform the type and location of the neural processes underlying self-motion perception.

ACKNOWLEDGMENTS

We gratefully thank Maria Liebsch, Reiner Boss, Michael Kerger and Harald Teufel for technical assistance and Florian Soyka for useful discussions.

GRANTS

AN, KAB and PP were supported by funds from the Max Planck Society. This work was also supported by the Brain Korea 21 PLUS Program through the National Research Foundation of Korea funded by the Ministry of Education. The funders had no role in study design, data collection and analysis, decision to publish, or preparation of the manuscript.

COPYRIGHT

5

HUMAN DISCRIMINATION OF HEAD-CENTRED VISUAL-INERTIAL YAW ROTATIONS

This chapter has been reproduced from an article submitted to Experimental Brain Research: Nesti A, Beykirch KA, Pretto P, Bülthoff HH. Human discrimination of head-centred visual-inertial yaw rotations. At the moment of publication of the present PhD thesis this article has been resubmitted to Experimental Brain Research after the first review phase.

5.1 ABSTRACT

To successfully perform daily activities such as maintaining posture or running, humans need to be sensitive to self-motion over a large range of motion intensities. Recent studies have shown that the human ability to discriminate self-motion in the presence of either inertial-only motion cues or visual-only motion cues is not constant but rather decreases with motion intensity. However, these results do not yet allow for a quantitative description of how self-motion is discriminated in the presence of combined visual and inertial cues, since little is known about visual-inertial perceptual integration and the resulting self-motion perception over a wide range of motion intensity. Here we investigate these two questions for head-centred yaw rotations (0.5 Hz) presented either in darkness or combined with visual cues (optical flow with limited lifetime dots). Participants discriminated a reference motion, repeated unchanged for every trial, from a comparison motion, iteratively adjusted in peak velocity so as to measure the participants' differential threshold, i.e., the smallest perceivable change in stimulus intensity. A total of 6 participants were tested at 4 reference velocities (15, 30, 45 and 60 deg/s). Results are integrated with previously published differential thresholds measured for visual-only yaw rotation cues using the same participants and procedure. Overall, differential thresholds increase with stimulus intensity following a trend described well by three power functions with exponents of 0.36, 0.62 and 0.49 for inertial, visual and visual-inertial stimuli, respectively. Despite the different exponents, differential thresholds do not depend on the type of sensory input significantly, suggesting that combining visual and inertial stimuli does not lead to improved discrimination performance over the investigated range of yaw rotations.

5.2 INTRODUCTION

When moving through the environment, humans need to constantly estimate their own motion for performing a variety of crucial tasks (e.g., maintaining posture in presence of external disturbances or controlling a vehicle). This estimate of self-motion, computed by the central nervous system (CNS), is the result of complex multisensory information processing of mainly visual and inertial cues and is inevitably affected by noise, and therefore uncertainty. This, for example, can cause two motions with different amplitudes to be perceived as similar, or can cause repetitions of the same motion to be perceived as different.

Over the last century, researchers have been investigating the properties of this perceptual variability, as well as its sources. While a large group of important studies focused on measuring the smallest perceivable motion intensity (absolute threshold) and its dependency on motion direction and frequency (cf. Guedry (1974)), only few studies addressed how the smallest perceivable *change* in motion intensity (differential threshold, DT) depends on the intensity of the supra-threshold motion (Zaichik et al. 1999; Mallery et al. 2010; Naseri and Grant 2012; Nesti et al. 2014a; Nesti et al. 2015). DTs for different intensities of combined visual and inertial motion cues have (to the best of our knowledge) not been investigated yet, as previous studies focused on how visual and inertial sensory cues independently contribute to the discrimination of self-motion. In this study, we investigate the human ability to discriminate rotations centred on the head-vertical axis (yaw) by measuring DTs for different supra-threshold motion intensities in the presence of congruent visual-inertial cues. Moreover, by comparing DTs for visual-inertial rotation cues with DTs for visual-only and inertial-only rotation cues (measured as three separate conditions), we address the question of whether redundant information from different sensory systems can improve discrimination of self-motion.

Supra-threshold motion discrimination

In everyday life, humans are frequently exposed to a wide range of self-motion intensities. For example during locomotion, head rotation velocities can range from 0 to 400 deg/s and even higher (Grossman et al. 1988). Recent studies investigated human DTs for different motion intensities (Zaichik et al. 1999; Mallery et al. 2010; Naseri and Grant 2012; Nesti et al. 2014a; Nesti et al. 2015). This is commonly done by presenting a participant with two consecutive motion stimuli and iteratively adjusting their difference in motion intensity until discrimination performance converges to a specific, statistically-derived level of accuracy (Gescheider 1997). By measuring DTs for different reference intensities, these studies showed that DTs increase for increasing motion intensities.

In three recent studies, Mallery et al. (2010), Naseri and Grant (2012) and Nesti et al. (2014a) measured human DTs for inertial-only motion cues (i.e., in darkness) for head-centred yaw rotations, forward-backward translations and vertical translations, respectively. Moreover, Nesti et al. (2015) measured DTs for yaw self-motion perception as evoked by a purely visual stimulation (vection). These studies have shown that DTs can be described well by a power function of the general form $\Delta S = k * S^{\,a}$, where ΔS is the DT, S is the stimulus intensity and k and a are free parameters that depend on the type of motion investigated. Of these two parameters, the exponent is the one that determines how fast DTs change with intensity: an exponent of 0 reflects DTs that do not depend on stimulus intensity whereas an exponent of 1 results in the well-known Weber's law (Gescheider 1988), which linearly relates DTs to stimulus intensity. In the studies mentioned above, the exponent ranges from 0.37 for yaw discrimination (Mallery et al. 2010) to 0.60 for discrimination of upward translations (Nesti et al. 2014a). Whether the functions describing DTs for visual-only and inertial-only stimuli also hold for congruent visual-inertial stimuli is still open question that we address with the present work.

Multisensory integration

In a natural setting, humans rely on visual, vestibular, auditory and somatosensory cues to estimate their orientation and self-motion. This information, coded by multiple sensory systems, must be integrated by the CNS to create a coherent and robust perception of self-motion. The theory of Maximum Likelihood Integration (MLI) provides a mathematical framework for how noisy sensory estimates might combine in a statistically optimal fashion (Ernst and Bülthoff 2004; Doya et al. 2007). In addition to providing a prediction of the multisensory percept, MLI theory also predicts the variance (i.e., the uncertainty) associated with that percept, based on the individual variances associated with each sensory modality. According to MLI, multisensory estimates always have lower variances than individual unisensory estimates (Ernst and Bülthoff 2004; Doya et al. 2007).

MLI is supported by a large amount of experimental evidence, for example in the fields of visual-auditory and visual-haptic integration (cf. Doya et al. (2007)). However, it is not unusual for psychophysical studies on visual-inertial integration to report deviations, sometimes substantial, from MLI predictions. For example, De Winkel et al. (2010) measured the human ability to estimate heading from visual, inertial, and congruent visual–inertial motion cues and observed that the variance associated with multimodal estimates was between the variances measured in the unisensory conditions. In a similar heading experiment, Butler et al. (2010) investigated human heading perception for visual and inertial stimuli as well as for congruent and incongruent visual-inertial stimuli. While congruent multisensory cues led to increased precision, for conflicting multisensory cues more weight was given to the inertial motion cue, resulting in multisensory estimates whose precision was not as high as MLI would predict. The MLI model was also rejected by de Winkel et al. (2013) in an experiment where participants discriminated between different yaw rotation intensities. In contrast, optimal or near-optimal integration of visual-inertial cues in heading tasks was reported in psychophysical experiments with humans (Butler et al. 2011; Prsa et al. 2012), as well as monkeys (Gu et al. 2008; Fetsch et al. 2009). Interestingly, Butler et al. (2011) suggested that stereo vision might be important in order to achieve MLI of visual and inertial cues, although results from Fetsch et al. (2009) contradict this hypothesis.

Overall, considering the high degree of similarity between experimental setups and procedures, such qualitative differences in results are surprising. A possible explanation could reside in the intrinsic ambiguity of visual stimuli (de Winkel et al. 2010), which contain information on both object motion and self-motion. Depending on properties of the visual stimuli, such as their duration, participants may or may not experience illusory self-motion perception (vection) (Dichgans and Brandt 1978). If vection is absent or incomplete, sensory integration is not expected to occur since the two visual and inertial sensory channels are believed to inform about two different physical stimuli: the motion of objects in the visual scene and self-motion.

Current study

The goal of this study is to psychophysically measure DTs for congruent visual-inertial yaw rotations over an intensity range of 15-60 deg/s and to identify the parameters of an analytical relationship (power function) that relates yaw DTs to motion intensity. Furthermore, we measure, in a separate condition and with the same participants, DTs to yaw rotation in darkness. We then compare DTs measured for inertial motion cues (inertial-only condition) and visual-inertial motion cues (visual-inertial condition) with DTs measured for visual motion cues (visual-only condition). The latter data were collected in a previous experiment on vection during constant visual yaw rotations conducted in our lab (Nesti et al. 2015). The present study is therefore designed to facilitate comparison of the data with Nesti et al. (2015) and to allow testing of an MLI model that predicts the variance of the bimodal (visual-inertial) estimate based on the variance of the unimodal (visual-only and inertial-only) estimates. Note that the use of constant visual rotation is a drastic deviation from the standard approaches described above for investigating MLI of visual-inertial cues and is motivated by the desire to ensure that, in the presence of visual-only cues, participants' discrimination is based on self-motion perception rather than object-motion perception. We hypothesise that DTs depend significantly on motion intensity and that providing visual-inertial motion results in DTs lower than those measured for unimodal motion cues, perhaps as low as MLI predicts.

This study extends current knowledge on self-motion perception by investigating motion discrimination with multisensory cues at different motion intensities. These types of stimuli occur frequently in everyday life and are therefore of interest to several applied fields. For instance, motion drive algorithms for motion simulators implement knowledge of self-motion perception to provide more realistic motion experiences within their limited workspace (Telban et al. 2005). Furthermore, models have been developed (Bos and Bles 2002; Newman et al. 2012) and employed to quantify pilots' perceptions of self-motion and orientation during both simulated and real flight, allowing estimation of any perceived deviation from reality in the simulator. By measuring DTs, we provide necessary information to adapt these multisensory models to account for the effect of stimulus intensity on the perception of self-motion, which in turn will result in more accurate predictions particularly at high motion intensities.

5.3 METHODS

Participants

Six participants (age 26 - 53, 1 female), 4 naïve and 2 experimenters (AN and KAB) took part in the study. They all had normal or corrected to normal vision, reported no history of balance or spinal disorders and no motion sickness susceptibility. Written informed consent was collected prior to the inclusion in the study, in accordance with the ethical standards specified by the 1964 Declaration of Helsinki.

Setup

The experiment was conducted using the MPI CyberMotion Simulator, an 8 degrees-of-freedom motion system capable of reproducing continuous head-centred yaw rotations (Figure 26, for technical details refer to Nieuwenhuizen and Bülthoff (2013); Robocoaster, KUKA Roboter GmbH, Germany). Participants sat inside the closed cabin in a chair with a 5-point harness and visual stimuli were presented on the white inner surface of the cabin door (approximately 60 cm in front of the participants' head) by means of two projectors (each 1920x1200 pixels resolution, 60 Hz frame rate). For this experiment, a field of view of approximately 70x90 deg and an actual stimulus resolution of approximately 20 pixels/deg were used. Participants wore headsets that played white noise during stimuli presentation to mask noise from the simulator motors and provide continuous communication with the experimenter (for safety reasons). The

Figure 26 *Experimental setup. Participants sat inside the simulator cabin and were presented with visual stimuli projected on the inner surface of the cabin door. The inset provides a picture of the visual stimulus.*

participant's head was restrained with a Velcro band, which combined with careful instruction to maintain an upright posture helped participants avoid Coriolis effects (Guedry and Benson 1976; Lackner and Graybiel 1984), i.e., the illusory perception of rolling/pitching following head tilts during constant velocity yaw rotations. Participants controlled the experiment with a button box with three active buttons; one was used to initiate the stimulus (control button) and the other two for providing a forced choice response (response buttons). As per instruction, the button box was held between the participants' knees, an active effort to help minimize proprioceptive information from the legs. The seat was also wrapped in foam to help mask vibrations of the simulator.

Stimuli

In both the visual-inertial and the inertial-only conditions, inertial stimuli consisted of 0.5 Hz sinusoidal yaw rotations centred on the participant's head. Each stimulus was composed of two consecutive parts characterized by two different peak amplitudes, a reference amplitude and a comparison amplitude, whose presentation order was randomized. The stimulus velocity first increased from 0 deg/s to the first peak amplitude following a raised half-cycle cosine mask of 1 second. This amplitude was then maintained for 5 seconds (2.5 cycles) before changing, again by means of a 1 second raised half-cycle cosine mask, to the comparison amplitude. After 5 seconds (2.5 cycles) the stimulus was terminated by decreasing its amplitude to 0 deg/s through a 3 second raised half-cycle cosine mask. The velocity profile of a typical stimulus is illustrated in Figure 27. Different stimulus onset and offset durations are used to hinder comparison of the two constant amplitudes based on stimulus accelerations. As shown by Mallery et al. (2010) through both modelling and experimental approaches, no confound is to be expected due to velocity storage for such stimuli, i.e., the perception of rotation that persists after the rotational stimulus stops (Bertolini et al. 2011). The stimuli designed for this study resemble those

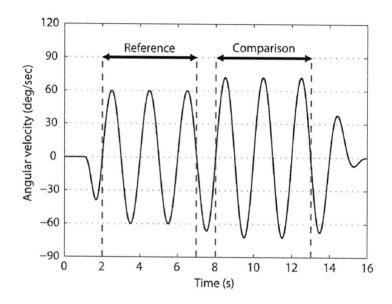

Figure 27 *Velocity profile of a typical stimulus composed of a reference amplitude of 60 deg/s and a comparison amplitude of 72 deg/s.*

employed by (Mallery et al. 2010) to the greatest possible extent to favour comparison of experimental findings.

Depending on the experimental condition, stimuli were always presented either in darkness (inertial condition) or combined with a virtual visual environment (visual-inertial condition) projected on the inner wall of the cabin (60 cm away from the participant). In the inertial condition projectors were off and participants were instructed to close their eyes. Visual stimuli, generated with authoring software for interactive 3D applications (Virtools, 3DVIA), consisted of limited lifetime dots (Figure 26) displayed on the surface of a virtual cylinder whose axis coincided with the head-vertical axis of the participants. The radius of the virtual cylinder (5 meters) was chosen to achieve a satisfactory visual appearance on the screen (i.e., texture resolution and object size). Dot life was set to 1 second to ensure that no dot outlived a full cycle of the sinusoidal motion, thereby preventing participants from comparing dots' travelled distances. The number of dots in the scene was maintained constant and the appearance delay was selected randomly between 0 and 200 ms. Each dot's diameter as it appeared on the inner wall of the cabin was 3 cm and remained constant for the entire lifetime of the dot. Visual and inertial sinusoidal rotations always had equal intensity and opposite direction, resulting in a congruent multisensory experience of self-motion, that is, the visual scene was perceived as earth-stationary. No visual fixation was used, thereby preserving the participants' natural behaviour.

Similar to (Nesti et al. 2015), participants were continuously rotating in each session around the head-vertical axis at the constant velocity of 20 deg/s. Although the perception of constant inertial rotations disappears within a few seconds after rotation onset (Bertolini et al. 2011), such motion generates vibrations (vibration rms of 0.08 m/s^2) unrelated to the stimulus, which serves multiple purposes. First, as suggested by Butler et al. (2010), when comparing reference and comparison stimuli, stimulus-unrelated vibrations could mask stimulus-related vibrations from the simulator, which are known to be amplitude dependent (Nesti et al.

2014b). Second, by setting the reference amplitude to 0 deg/s, it is possible to measure the yaw absolute threshold in a discrimination task, as it prevents participants from merely performing a vibration detection task (Mallery et al. 2010; Merfeld 2011). Finally, this allows for a more direct comparison with DTs estimated by (Nesti et al. 2015). The direction of the constant rotation was reversed approximately every 15 minutes and stimulus presentation began 1 minute after constant velocity was reached to guarantee disappearance of rotational motion perception.

An inertial measurement unit (YEI 3-Space Sensor, 500 Hz) mounted on top of a participant's head was used to verify the absence of centripetal accelerations during constant and sinusoidal yaw rotations and for measuring temporal disparities between visual and inertial motion, a common concern for mechanical and visual systems. This procedure revealed that, when commanded simultaneously, the visual motion preceded the physical motion by approximately 32 ms. Because increasing temporal disparities diminish the influence that multimodal cues have on each other (van Wassenhove et al. 2007; van Atteveldt et al. 2007), temporal disparities were minimized by delaying visual stimuli by 2 frames, which corresponds to approximately 33 ms at the projectors frame rate of 60 Hz.

Similar to Nesti et al. (submitted), participants were continuously rotating in each session around the head-vertical axis at the constant velocity of 20 deg/s. Although the perception of constant inertial rotations disappears within a few seconds after rotation onset (Bertolini et al. 2011), such motion generates vibrations (vibration rms of 0.08 m/s^2) unrelated to the stimulus which serve multiple purposes. First, as suggested by Butler et al. (2010), stimulus-unrelated vibrations could mask stimulus-related vibrations from the simulator, which are known to be amplitude dependent (Nesti et al. 2014b). Second, by setting the reference amplitude to 0 deg/s it is possible to measure the yaw absolute threshold in a discrimination task as it prevents participants from merely performing a vibration detection task (Mallery et al. 2010; Merfeld 2011). Finally, this allows for a more direct comparison with DTs estimated by Nesti et al. (submitted). The direction of the constant rotation was reversed approximately every 15 minutes and stimulus presentation began 1 minute after constant velocity was reached to guarantee disappearance of rotational motion perception.

An inertial measurement unit (YEI 3-Space Sensor, 500 Hz) mounted on top of a participant's head was used to verify the absence of centripetal accelerations during constant and sinusoidal yaw rotations and for measuring temporal disparities between visual and inertial motion. When commanded simultaneously, the visual preceded the physical motion by approximately 32 ms. Because increasing temporal disparities diminish the influence that multimodal cues have on each other (van Wassenhove et al. 2007; van Atteveldt et al. 2007), temporal disparities were minimized by delaying visual stimuli by 2 frames, which corresponds to approximately 33 ms at the projectors frame rate of 60 Hz.

Procedure

Before stimulus presentation, participants sat in darkness (inertial condition) or in front of the visual environment, initially stationary with respect to the participants (visual-inertial condition). Stimuli were initiated by the participants through the button box and started 1 second after the control button was pressed. A 5 second tone accompanied the presentation of both the reference and the comparison amplitudes. After hearing a beep indicating the end of the stimulus, participants were asked "which rotation felt stronger (1st or 2nd)?". Participants were specifically instructed to refer to the motion they felt during the two 5-seconds tone presentations and not during any other part of the stimulus. After a feedback beep, confirming that the answer was recorded, participants waited for 3 seconds before a beep signalled they could start the next stimulus. In the visual-inertial condition, the visual scene remained visible and stationary with respect to the participants during the time between stimuli.

Both the inertial and the visual-inertial conditions were divided into 4 sessions of approximately 45 minutes each, with a 10 minute break roughly in the middle of the session to avoid fatigue. Each participant was only allowed to complete 1 session per day. In every session, the participant's DT was measured for one of four reference velocities (15, 30, 45 or 60 deg/s) using a psychophysical two-interval forced-choice (2IFC) procedure. While the reference velocity remained constant throughout the whole session, comparison velocities were adjusted for every trial according to an adaptive staircase algorithm: the stimulus level was decreased after 3 consecutive correct responses and increased after every incorrect response (3-down 1-up rule (Levitt 1971)). Such an algorithm converges where the probability of a single correct answer is 0.794 (cube root of 0.5), i.e., when the probability of a stimulus increase (wrong answer) or decrease (3 consecutive correct answers) is equal (p=0.5). The comparison velocity c_0 for the first trial was obtained by multiplying the reference velocity by 1.2. The step size, initially set at 2 deg/s, was halved every 5 reversals. Sessions were terminated after 13 reversals (final step size of 0.5 deg/s). Typical staircases for one participant are illustrated in Figure 28. All participants completed the inertial-only condition before the visual-inertial condition. Reference velocities were tested in random order. An additional session was run to measure the yaw absolute threshold (reference velocity set to 0 deg/s) for inertial-only motion stimuli. In this session the initial comparison velocity was set to 2 deg/s with a constant step size of 0.1 deg/s.

Visual condition

Human discrimination of yaw rotations in the presence of visual cues alone was investigated previously by Nesti et al. (2015), to allow comparison with inertial and visual-inertial cues, as was done in this work. Briefly, in Nesti et al. (2015) we measured DTs for circular vection for the same 6 participants of the present study and for the same 4 reference rotational velocities (15, 30, 45 and 60 deg/s). The study also employed the same setup and experimental procedure (2IFC, 3-down 1-up adaptive staircase). Visual

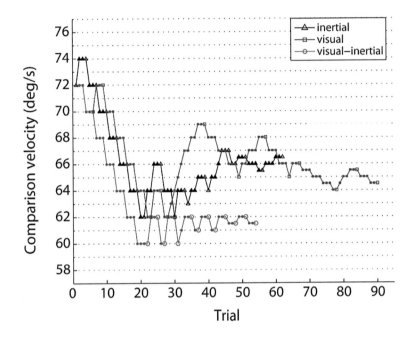

Figure 28 *Evolution of the adaptive algorithms for one participant in the inertial (black line), and visual-inertial (red line) conditions. Blue line represents data re-plotted from Nesti et al. (2015), where DTs for visual-only motion cues were measured using an identical adaptive procedure. Reference velocity was 60 deg/s. Empty markers indicate reversals.*

rotations were presented at constant velocity, a stimulus that is known to induce a compelling self-motion perception due to its lack of conflict between visual and inertial information (Dichgans and Brandt 1978). Indeed, human perception of head-centred constant rotations is hindered by the lack of angular accelerations. This results in non-conflicting visual and inertial sensory information during constant visual rotations irrespective from the intensity of the inertial rotation. To guarantee that a compelling self-motion perception was induced in the participants *at every trial*, visual rotations were terminated by participants via a button press only after the visual scene was confidently perceived as stationary, i.e., all the visual motion was attributed to self-motion. Note that this constitutes a qualitative difference with most of the published studies on MLI of visual-inertial cues in self-motion perception, where stimuli for the visual-only condition are obtained by simply removing the inertial component from the stimuli of the visual-inertial condition (Butler et al. 2010; de Winkel et al. 2010; Butler et al. 2011; Prsa et al. 2012; de Winkel et al. 2013). These arrangements for the measurement of visual-only DTs have the benefit of avoiding comparison of a self-motion percept (evoked in the inertial-only and in the visual-inertial condition) with an object motion percept, which most likely occurs when stationary participants are briefly exposed to visual objects moving on a screen with changing velocity. Although the stimuli from Nesti et al. (2015) differ from the stimuli employed in the present study in terms of stimulus frequency and visual environment, we argue that these differences do not hinder a meaningful comparison of the results of these two studies (see "Validity of study comparison" in Discussion). A combined analysis allows for comparison of yaw discrimination in response to visual-only or inertial-only cues. Moreover, it allows for investigation of how redundant sensory information from the visual and inertial sensory systems combine in the presence of multisensory motion cues.

Data analysis

For every condition, the last eight reversals of the staircase algorithm were averaged in order to compute the DT corresponding to the reference velocity and sensory modality tested. The DTs for each amplitude were averaged across participants for each of the three conditions, inertial-only, visual-inertial and visual-only (Nesti et al. 2015). The averages were fit for each condition to a power function of the form:

$$\Delta S = k * S^a \qquad (10)$$

where ΔS is the differential threshold and S is the stimulus intensity and k and a are free parameters. The choice of the power function is motivated by previous studies showing that the power function provides a good description of DTs for self-motion perception as well as for other perceptual modalities (Guilford 1932; Mallery et al. 2010; Nesti et al. 2014a; Nesti et al. 2015).

A repeated-measures analysis of the covariance (rmANCOVA) was run to assess the effect of the factor "condition" (3 levels: "inertial", "visual" and "visual-inertial") and of the covariate "motion intensity". In order to perform the rmANCOVA using the power function model, the following transformation of the data was required:

$$\log(\Delta S) = \log(k) + a * \log(S) \qquad (11)$$

Additionally, to assess whether the integration of visual and inertial cues followed the MLI model in this experiment, an rmANCOVA was run to compare participants' DTs in the visual-inertial condition with MLI predictions based on their own DTs as measured in the visual-only and inertial-only conditions. Note that it is common practice to test MLI using the variance of the physiological noise underlying the decision process rather than the experimentally derived thresholds (see e.g., Butler et al. 2010; de Winkel et al. 2013). For a two interval discrimination task, such as the one employed here, this requires dividing the DTs by 0.58 (Merfeld 2011). Such a linear transformation of the data does not however affect the results of the statistical analysis and we therefore test MLI directly on the measured DTs using the following equation (Ernst and Bülthoff 2004):

$$\overline{DT_{vi}}^2 = \frac{DT_v^2 * DT_i^2}{DT_v^2 + DT_i^2} \qquad (12)$$

where DT_v and DT_i are the DTs measured in the visual-only and inertial-only conditions, respectively, for every reference intensity and $\overline{DT_{vi}}$ is the MLI prediction for the DT at the given reference velocity in the visual-inertial condition.

Stimulus noise analysis

When reproducing motion commands, motion simulators inevitably introduce noise that affects the amplitude and spectral content of the intended inertial stimulus. As extensively discussed in Nesti et al. (2014b), analysing the noise introduced in the stimulus by the simulator provides important insights in the study of self-motion perception, as it allows dissociation of the mechanical noise of the experimental setup from the noise that is inherent in the perceptual processes. A Signal to Noise Ratio (SNR) analysis (Nesti et al. 2014b) of the motion stimuli was therefore conducted using an inertial measurement unit (STIM300 IMU, Sensonar AS, 250 Hz) rigidly mounted on the floor of the simulator cabin. The SNR expresses the relative amount of commanded signal with respect to motion noise and is therefore an indicator of similarity between commanded and reproduced motion. For every reference velocity, 20 stimulus repetitions were recorded and the noise was then extracted by removing the motion command from the recorded signal (Nesti et al. 2014b). Average SNRs were computed for every reference stimulus and tested by means of an ANCOVA to investigate the effect of motion intensity on the motion SNRs.

$$SNR = \left(\frac{rms_{signal}}{rms_{noise}}\right)^2 \qquad (13)$$

where *rms* stands for the root mean square of the noise signal and of the recorded signal (Nesti et al. 2014b).

5.4 RESULTS

Motion analysis of the reference stimuli, illustrated in Figure 29, shows a significant increase in stimulus SNR for increasing amplitudes of the velocity command ($F(1,78)=113.8$, $p<0.001$). This is a common feature of motion simulators (cf. Nesti et al. 2014b) and is expected to facilitate motion discrimination of higher as compared to lower motion intensities for those perceptual systems (including the human perceptual system (Greig 1988)) whose discrimination performances increase with SNRs. The fact that human DTs for self-motion increase for increasing motion intensities (Mallery et al. 2010; Naseri and Grant 2012; Nesti et al. 2014a, present study) could indicate an additional noise source inherent to the perceptual system and proportional to stimulus intensity.

During the experiment, each condition took approximately 40 minutes and required on average 61 trials. No session needed to be terminated because of fatigue or other reasons and no participant reported symptoms of motion sickness.

The absolute threshold measured in the inertial-only condition was 0.87 ± 0.13 deg/s, a value that is consistent with previous studies (see e.g., Zaichik et al. 1999; Mallery et al. 2010; Valko et al. 2012; Roditi and Crane 2012).

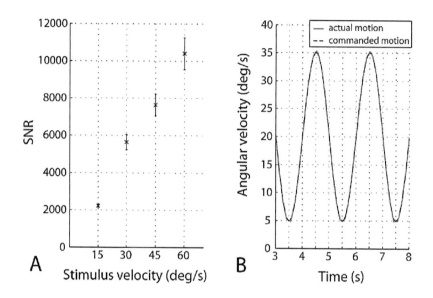

Figure 29 *SNR analysis of the motion stimuli employed in this study. A SNRs increase for increasing rotational intensities, resulting in the highest comparison (peak amplitude of 60 deg/s) having an SNR approximately 5 times higher than the lowest comparison (peak amplitude of 15 deg/s). Error bars represent ±1 SEM. B Comparison of the commanded and recorded motion profile for one stimluus with 15 deg/s amplitude.*

Fitting equation 1 (power function) to inertial, visual and visual-inertial DTs averaged for each reference velocity results in gain coefficients k_i, k_v and k_{vi} of 1.33, 0.55 and 0.76 and in exponent coefficients a_i, a_v and a_{vi} of 0.36, 0.62 and 0.49, where the subscripts i, v and vi stand for inertial, visual and visual-inertial, respectively (Figure 30). Goodness of fit, quantified by the R^2 coefficient, are 0.88, 0.89 and 0.99, respectively. Note that the inertial condition replicates the findings of Mallery et al. (2010) remarkably well, where k_i = 0.88 and a_i = 0.37. The overall higher thresholds found in our study, reflected in the higher gain (1.33 vs 0.88), are likely due to the use of a different simulator. However, similar exponents indicate that the effect of motion intensity on self-motion discrimination in darkness is qualitatively consistent between studies. This is only partially surprising given the high level of similarity in the experimental methods. A linear fit resulted in intercept coefficients q_i, q_v and q_{vi} of 2.88, 1.73 and 2.05 and in slope coefficients m_i, m_v and m_{vi} of 0.05, 0.09 and 0.06. R^2 coefficients are 0.91, 0.87 and 0.99 in the inertial-only, visual-only and visual-inertial condition, respectively. Although the linear model provides a slightly superior fit than the power function model for the inertial-only condition, we performed the rmANCOVA using the power function model as it should generalize better for larger ranges of sensory input amplitudes (Guilford 1932; Teghtsoonian 1971).

The ANCOVA revealed that DTs increased significantly with motion intensity ($F(1,63)=32.55$, $p<.001$), confirming previous results on self-motion discrimination in the presence of visual-only or inertial-only cues and extending the analysis to the case of visual-inertial cues. However, DTs did not depend on the cue type ($F(2,63)=1.59$, $p=0.21$), i.e. whether participants experienced inertial, visual or visual-inertial stimuli. Predictions based on MLI were contradicted by measured visual-inertial DTs (Figure 31), with significantly higher results ($F(1,40)=5.93$, $p=0.02$).

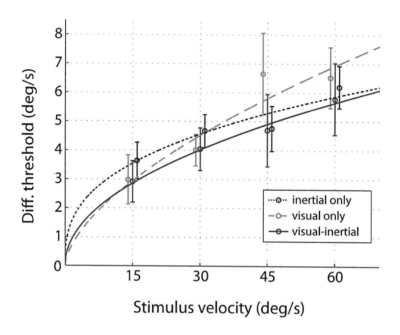

Figure 30 *DTs for yaw rotations in presence of inertial (blue), visual (green) and visual-inertial (red) motion cues are well described by three power laws. DTs for visual cues are re-plotted from Nesti et al. (2015). Error bars represent ±1 SEM.*

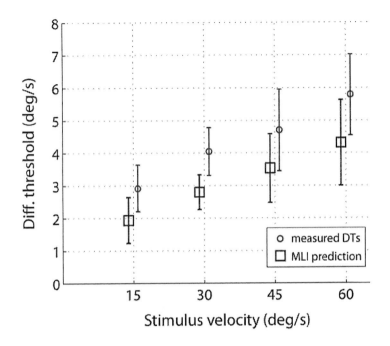

Figure 31 *Comparison between measured DTs (circles) and predicted DTs (squares) based on MLI. Data do not support models of statistically optimal integration of visual and inertial sensory information. Error bars represent ±1 SEM.*

5.5 DISCUSSION

Human self-motion perception involves the contribution of different sensory information from the visual, vestibular, auditory and somatosensory systems. In this study, we investigated human discrimination of self-motion for a wide intensity range of yaw rotations in darkness (inertial-only motion cues) and with congruent visual-inertial motion cues. Measured DTs increase with motion intensity following a trend described well by a power function, in agreement with previous studies on rotations and translations in darkness (Mallery et al. 2010; Naseri and Grant 2012; Nesti et al. 2014a) and for visually induced self-motion perception (Nesti et al. 2015). The use of a power function is consistent with previous work on self-motion perception (Mallery et al. 2010; Nesti et al. 2014a; Nesti et al. 2015) and resulted in a high goodness-of-fit. Note however that a Weber's law fit also provides a similar goodness-of-fit.

In the next sections the relationship between DTs and motion intensity and the sub-optimal integration that emerged from the present study are discussed in detail.

Discrimination of yaw rotations

Human DTs for self-motion are not independent from the intensity of the motion (as would be the case for a linear system with constant noise) but rather increase for increasing motion intensities. The present study shows that such behaviour is present not only for visual-only (Nesti et al. 2015) and inertial-only conditions (Zaichik et al. 1999; Mallery et al. 2010; Naseri and Grant 2011; Nesti et al. 2014a, present study), but is encountered also for congruent visual and inertial sensory cues. This indicates that the perceptual processes converting physical to perceived motion are nonlinear and/or affected by stimulus-dependent noise (with the amount of noise increasing with the intensity of the physical stimulus). In contrast, responses to head rotations are linear with constant inter-trial variability for neurons in the vestibular afferents and vestibular nuclei (Sadeghi et al. 2007; Massot et al. 2011), as well as for eye movements (Pulaski et al. 1981; Weber et al. 2008). A comparison between psychophysical and physiological studies suggests therefore that nonlinearities and/or stimulus-dependent increases in physiological noise occur further along the neuronal pathways processing sensory information and are likely due to central processes, multisensory integration mechanisms and/or cognitive factors. It is however interesting to note that in neural recordings from the vestibular nuclei of macaque monkeys increased variability was observed for faster compared to slower inertial (Massot et al. 2011), visual (Waespe and Henn 1977) and visual-inertial (Allum et al. 1976) yaw rotational cues. We hypothesize that this increase in variability reduces discrimination performances at high stimulus velocities. Future studies are required to better quantify the relationship between stimulus intensity, neural activity and behavioural responses.

Stimulus-dependent DTs might also represent an efficient strategy of the CNS to account for how frequently a particular motion intensity occurs in everyday life. This would indeed result in smaller DTs for low rotation intensities, as they are more common than large rotations during everyday experience. To better illustrate this concept we present in Figure 32 rotational velocity intensities recorded with an inertial sensor (YEI 3-Space Sensor, 500 Hz) over 40 minutes of normal activity (running) and fit with an exponential distribution. A simple model with 2 parameters, gain and offset, is able to describe the increasing trend of DTs well.

Note that the simple model from Figure 32 only serves as an illustrative example. A more systematic approach for using stimulus statistics to model perceptual responses is presented in Wei and Stocker (2013).

Multisensory integration

In this study we investigated multisensory integration in a yaw intensity discrimination task by comparing DTs for inertial-only and visual-only

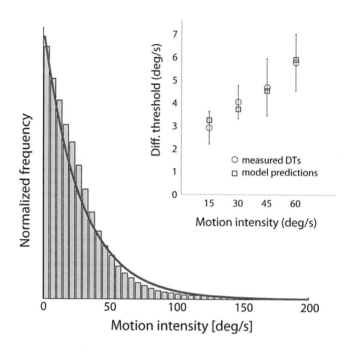

Figure 32 *Physical stimulus statistics obtained using an IMU are presented in a histogram where bars represent the normalized frequency of yaw rotational velocity during a 40 minutes running session. Normalized frequencies are obtained by dividing the histogram of yaw data samples by its area. Fitting data with an exponential distribution (red line, y(S) = 28.5 * exp(-28.5 * S), where S is the stimulus intensity and y(S) is the exponential distribution) allows developing of a simple model (ΔS = a + b * 1 / y(S)) that relates DTs to motion intensity by accounting for how frequently a particular intensity occurs. Error bars represent ±1 SEM.*

motion stimuli with DTs for congruent (i.e., redundant) visual-inertial cues. Although a number of studies indicated MLI as a valid model of visual-inertial cue integration (see e.g., Gu et al. 2008; Fetsch et al. 2009; Butler et al. 2011; Prsa et al. 2012), our data do not seem to follow MLI. This is only partially surprising, as we are not the first to report substantial deviations from MLI (Telford et al. 1995; Butler et al. 2010; de Winkel et al. 2010; de Winkel et al. 2013). However, when comparing this study with the existing literature, it is important to consider two main differences. First, the great majority of visual-inertial integration studies used a heading task, rather than a rotation intensity discrimination task as we have. Although MLI has been suggested as a general strategy for multisensory integration, the stimuli are radically different and even involve different vestibular sensors (note that a heading stimulus is composed by linear translations only), therefore caution is advised in the generalization of the results. The only other studies, of which we are aware, that employed yaw stimuli are from Prsa et al. (2012), whose findings support MLI, and from De Winkel et al. (2013), where the MLI model is rejected. Second, the stimuli we chose for testing for MLI were designed to avoid visual-inertial conflicts. This required an inertial-only stimulus to which the visual system is insensitive (i.e., motion in darkness) and a visual-only stimulus to which the inertial systems are insensitive (i.e., rotation at constant velocity (Nesti et al. 2015), which lacks inertial accelerations). To the best of our knowledge, such stimuli have not been previously employed for validating MLI of visual-inertial motion cues. Instead, perceptual thresholds for visual-only cues were always investigated by removing the inertial component from the visual-inertial stimulus, a choice that minimizes experimental manipulations but might lead to visual-inertial sensory conflicts.

In light of our experimental results, the visual-inertial DTs may be reconciled with MLI through the theory of causal inference (Beierholm et al. 2008; Shams and Beierholm 2010), which predicts that sensory integration is subordinate to whether stimuli are perceived as originating from the same physical event or not. Although in the present study, the visual and inertial stimuli were always congruent in representing head centred rotations, we have to consider the possibility that they were not always perceived as congruent by the participants. Indeed, the simple fact that visual stimuli were computer-generated virtual objects might induce in the participants expectations of incongruence with the actual motion (the visual and inertial stimuli "belong" to different environments). Causal inference theory suggests that in this event stimuli are segregated and participants respond based on the information coming from either one of the 2 sensory channels. Statistical models, other than causal inference, have been suggested in the literature to account for the possibility that stimuli are not integrated according to MLI because they are perceived as incongruent (see De Winkel et al. (2013) for a review). For instance, a "switching strategy" model could be applied to our data by assuming that stimuli perceived as congruent are integrated according to MLI whereas stimuli perceived as incongruent are segregated and the response is based only on one sensory modality (e.g., the inertial). Equation 3 would then be modified as follow:

$$\overline{\sigma_{iv}}^2 = \frac{\sigma_i^2 * \sigma_v^2}{\sigma_i^2 + \sigma_v^2} * \pi + \sigma_i^2 * (1 - \pi) \tag{14}$$

leading to an estimated average probability (π) of 0.33 that participants perceived the stimuli as congruent.

Nonlinear self-motion perception models

Human self-motion perception models compute how people update the estimate of their motion in space in response to physical motion. Several models were developed combining knowledge of sensor dynamics, oculomotor responses, psychophysics and neurophysiology (Merfeld et al. 1993; Bos and Bles 2002; Zupan et al. 2002; Newman et al. 2012). Despite capturing a large variety of perceptual phenomena well, to the best of our knowledge no published model can account for the decrease in discrimination performances with increasing motion intensity. The experimental data collected in this study and by Nesti et al. (2015) constitutes a crucial step towards a more complete approach to self-motion perception models. Considering that the DTs measured here increases with stimulus intensity and were not affected by manipulation of the type of sensory information, a natural and straightforward choice would be to implement a single, common nonlinear process after the integration of the visual and inertial sensory pathways. Future studies should be dedicated to measuring rotational and translational multisensory DTs for the remaining degrees of freedom, implementing perceptual nonlinearities in computational models of human self-motion perception and validating these models using alternative motion profiles and experimental paradigms (e.g., Maximum Likelihood Difference Scaling, Maloney and Yang 2003).

Validity of study comparison

We compared our results with DTs for vection (Nesti et al. 2015) to test the hypothesis that redundant information from the visual and inertial sensory systems is perceptually combined in a statistically optimal fashion. Comparison of DTs measured here and in Nesti et al. (2015) is particularly natural because of the high similarity between the studies (experimental setup, participants, procedure and stimulus intensities). However, two important differences should be discussed.

First, in the present study 0.5 Hz sinusoidal motion profiles were used, whereas in Nesti et al. (2015) we measured vection DTs for constant (0 Hz) yaw rotations and stimuli were self-terminated by the participant to account for the high individual variability in vection onset time (Dichgans and Brandt 1978). These choices were made in order to measure DTs for stimuli as free of visual-inertial conflicts as possible, ensuring that all the visual motion is attributed to self-motion rather than object motion. Note how, for supra-threshold motion intensities, a visual stimulus at 0.5 Hz combined with no inertial motion will surely evoke a visual-inertial sensory conflict, as the continuous changes in the velocity of the visual environment conflicts with the lack of acceleration signal from the inertial sensory systems. Evidence that conflicts between visual and inertial cues could confound self-motion perception is provided for instance by Johnson et al. (1999), who showed that in bilateral labyrinthectomized patients, who lack one of the main sources of inertial information (i.e., the vestibular system), vection latencies are shorter than those of healthy subjects. Comparing DTs for constant rotations with DTs for visual-inertial rotations at 0.5 Hz requires however the assumption that visual responses remain constants within this frequency range. Previous studies indicates that postural, psychophysical and neurophysiological responses to visually simulated self-motion show low-pass characteristics (Robinson 1977; Mergner and Becker 1990; Duh et al. 2004). For instance, visual responses in the vestibular nuclei only begin to attenuate for frequencies higher than 0.03 Hz (Robinson 1977), while subjective reports of circular vection intensities remain approximately constant for frequencies between 0.025 and 0.8 Hz (Mergner and Becker 1990). It is however reasonable to expect that this attenuation is at least in part due to multisensory conflicts that arise at stimulus frequencies to which the inertial sensors respond. Further studies in labyrinthectomized patients might help in clarifying the dependency of visual responses on frequency, although it should not be forgotten that the vestibular system is not the only system contributing to self-motion perception.

The second important difference involves the different visual stimulus: whereas in the present study we employed a limited lifetime dot field, in Nesti et al. (2015) we employed a 360 degrees panoramic picture of a forest. Although it is known that different visual environments (e.g., with different spatial frequencies) affect vection onset time (Dichgans and Brandt 1978), we suggest that DTs after vection arose (i.e., when the visual environment is perceived as stationary) depends only on the velocity of the optic flow and not on the texture of the visual stimulus. This difference could be obviously eliminated in future studies by employing the same virtual environment for every condition and ensuring that it does not provide visual references.

To the best of our knowledge, this is the first study that focuses on minimizing sensory conflicts when testing MLI of visual-inertial cues for self-motion perception. While differences between the visual stimuli in the visual-only and visual-inertial conditions advise for caution in the interpretation of the results, we believe that preventing confounds between object-motion and self-motion perception in psychophysical experiments is an important step towards the understanding of the perceptual processes underlying the integration of visual-inertial cues.

ACKNOWLEDGMENTS

We gratefully thank Maria Lächele, Reiner Boss, Michael Kerger and Harald Teufel for technical assistance and Mikhail Katliar for useful discussions.

GRANTS

This work was supported by the Brain Korea 21 PLUS Program through the National Research Foundation of Korea funded by the Ministry of Education. The funders had no role in study design, data collection and analysis, decision to publish, or preparation of the manuscript.

6

VARIABLE ROLL-RATE PERCEPTION IN DRIVING SIMULATION

This chapter has been reproduced from an article submitted for publication in The Society for Modeling & Simulation International, special issue: Driving Simulation. Part of the work described in this chapter is also presented in the proceedings of the Driving Simulation Conference 2014: Pretto P, Nesti A, Nooij S, Losert M, Bülthoff HH (2014) Variable roll-rate perception in driving simulation. Driving Simulation Conference, Paris, 2014.

6.1 ABSTRACT

In driving simulation, simulator tilt is used to reproduce linear acceleration. In order to feel realistic, this tilt is performed at a rate below the tilt-rate detection threshold, which is usually assumed constant. However, it is known that many factors affect the threshold, like visual information, simulator motion in additional directions, or active vehicle control. In the present study we investigated the effect of these factors on roll-rate detection threshold during simulated curve driving. Ten participants reported whether they detected roll in multiple trials on a driving simulator. Roll-rate detection thresholds were measured under four conditions. In the first three conditions, participants were moved passively through a curve with: (i) roll only in darkness; (ii) combined roll/sway in darkness; (iii) combined roll/sway and visual information. In the fourth condition participants actively drove through the curve. Results showed that roll-rate perception in vehicle simulation is affected by the presence of motion in additional directions. Moreover, an active control task seems to increase the detection threshold, i.e. impair motion sensitivity, but with large individual differences. We hypothesize that this is related to the level of immersion during the task.

6.2 INTRODUCTION

In dynamic vehicle simulation, motion cueing algorithms (MCAs) aim to adapt the original vehicle motion to the limited capabilities of simulators, while preserving at the same time the perceptual realism of the simulation. The goal of MCAs is therefore to transform the linear and angular accelerations of the simulated vehicle into translations and rotations of the motion platform, such that perceptually equivalent specific forces and rotations are provided to the driver.

Most MCAs are based on washout filters (Nahon and Reid 1990), which split the input linear accelerations into high-frequency and low-frequency components. The high-frequency components are integrated to produce the translational motion of the platform, while the low-frequency components are reproduced by tilting the platform. The tilt of the platform is used by MCAs to simulate sustained accelerations (otherwise not reproducible) exploiting the so-called tilt-coordination technique (Nahon and Reid 1990; Groen and Bles 2004). This is one of the most used "perceptual tricks" in motion cueing, which relies on the tilt-translation ambiguity (Einstein 1908; Angelaki and Yakusheva 2009). Indeed, under certain conditions the simulator tilt can be perceived as linear acceleration, as the reorientation of the body with respect to gravity causes the sensation of being forced into (or away from) the seat. This illusion occurs because different combinations of linear accelerations and static body tilt result in similar gravito-inertial forces acting on the humans inertial sensory systems (primarily vestibular and somatosensory). This is particularly effective when concurrent translational motion is visually presented (Groen and Bles 2004; Berger et al. 2007). The tilt-coordination technique exploits the inability of humans to resolve the tilt-translation ambiguity, and uses simulator tilt to induce the illusory perception of sustained linear acceleration.

However, the illusion is spoiled if the platform tilt is detected by the driver (Stratulat et al. 2011). This happens when the tilt velocity exceeds the perceptual threshold, inducing the sensation of rotational motion and resolving the ambiguity. Therefore, to preserve the realism of the simulation, a rate limiter saturates the platform tilt-rate below perceptual threshold. A commonly used saturation value is 3 deg/s (Groen and Bles 2004).

Human tilt-rate detection threshold is usually measured in darkness (Zaichik et al. 1999; Heerspink et al. 2005; Valko et al. 2012). Yet, it is known that motion perception thresholds can vary in the presence of visual information (Groen and Bles 2004; Valente Pais et al. 2006), simulator motion in additional directions, i.e. increased motion complexity (Zaichik et al. 1999), active vehicle control (Hosman and van der Vaart 1978; Vroome et al. 2009; Nesti et al. 2012), or even cognitive expectations (Wertheim et al. 2001). All these factors are actually present in a typical driving simulation. Still, most of current MCAs assume constant tilt rate thresholds, often derived from studies where simple motion stimuli (e.g. purely sinusoidal acceleration profiles in darkness) were investigated. Therefore, a better understanding of how motion complexity, visual information and active control affect the perception of simulator motion may help in improving the efficiency of tilt-coordination techniques.

In this study, we investigated for the first time in the same experiment (using the same simulator and methodology for all experimental conditions), the effect of each of these factors on roll-rate detection threshold during simulated curve driving.

6.3 METHODS

Setup

The experiment was conducted on the CyberMotion Simulator (CMS) at the Max Planck Institute for Biological Cybernetics. The CMS was developed as an alternative to traditional dynamic simulators based on hexapod systems (Nieuwenhuizen and Bülthoff 2013). It is a 8-dof serial robot, where a 6-axes industrial robot manipulator is mounted on a linear rail and equipped with a motorized cabin at the end effector (Figure 33, top). The cabin is equipped with a stereo projection system and mounting possibilities for haptic control devices used for flight and driving simulation (Figure 33, bottom). In the driving configuration it is equipped with force-feedback steering wheel (Sensodrive GmbH, Germany) and pedals, and a large projection screen (160 x 90 deg FoV) with two WUXGA (1920x1200 pixels) projectors. For this study, the motion was generated using a classical washout filter, adapted to the cylindrical workspace of the CMS (Robuffo Giordano et al. 2010b). No linear rail was used and the lateral motion was mapped into a circular trajectory (Nesti et al. 2012). The vehicle dynamics and the visualization environment were provided by the simulation software CarSim (Mechanical Simulation, Michigan, US). The visual scene resembled a flat skidpad (Figure 33, bottom), and no roll-motion was present other than the one originating from the car suspensions.

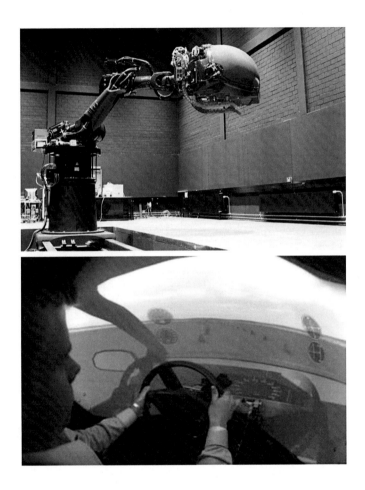

Figure 33 *The Max Planck Institute CyberMotion Simulator: exteriors (top) and cabin interior (bottom).*

Experimental manipulations

The rate limiter of the tilt (roll) channel of the washout filter was manipulated during this experiment. Roll-rate detection thresholds were estimated under four conditions:

— "Roll": roll only in darkness;
— "+Sway": combined roll/sway in darkness;
— "+Visual": combined roll/sway and visual information whilst passively moved through a curve;
— "+Active": combined roll/sway and visual information whilst actively driving around a curve.

An overview of the experimental conditions is provided in Table 5.

	"Roll"	"+Sway"	"+Visual"	"+Active"
Roll	Present	Present	Present	Present
Sway	Absent	Present	Present	Present
Visual	Absent	Absent	Present	Present
Active	Absent	Absent	Absent	Present

Table 5 *Experimental conditions. The first row reports the labels assigned to every condition.*

Procedure

Ten participants (three females), aged between 25 and 36 (mean = 29, SD = 3.5) took part in the experiment. All had a valid driving license for at least three years and self-reported regular car usage. The experiment was divided into four sessions, over different days. Each session started with three practice trials to familiarize the participant with the task. When the participant initiated a trial by button press, the car accelerated automatically on a straight road until a constant speed of 70 km/h was reached. The speed was then maintained constant throughout the whole trajectory. During the acceleration phase, the surround scene and the layout of the curve were visible in all conditions (Figure 33, bottom). Before entering the curve section, the screen turned to black in conditions "Roll" and "+Sway"; while in conditions "+Visual" and "+Active" the outside view remained visible. The car progressed through the curve automatically (conditions "Roll", "+Sway", and "+Visual") or with the heading actively controlled by the participant (condition "+Active"). At the end of the curve the road was straight again (no active control required) and the following question appeared on the screen: "Did the car tilt (Y/N)?" The participant indicated the answer by pressing a button accordingly. When the answer was given the car decelerated and the simulator was brought back to the starting position for the next trial.

Thresholds were measured by iteratively adjusting the roll-rate saturation value according to the Single Interval Adjustment Matrix (SIAM) procedure (Kaernbach 1990; Shepherd et al. 2011). In 50% of the trials, the tilt coordination channel of the MCA was active (roll motion present), while in the other 50% tilt coordination was disabled (roll motion absent). Additional roll-motion of the car (e.g. suspensions) was not cued. The trials were randomly interleaved. The participants had to correctly identify whether the roll was present by answering the question above. The adjustment matrix of the four possible outcomes was set up to induce a neutral response criterion: the answer "yes" in presence of roll (hit) decreased the roll-rate saturation value for the next trial of one step size; the answer "no" in absence of roll (correct rejection) left the roll value unaltered; the answer "yes" in absence of roll (false alarm) increased the next roll value of two step sizes; the answer "no" in presence of roll (miss) increased the roll rate of one step size for the next trial. The SIAM is given in short form in Table 6.

Motion Stimulus	Answer: "Yes"	Answer: "No"
Roll	Hit [-1]	Miss [+1]
No Roll	False Alarm [+2]	Correct Rejection [0]

Table 6 *The SIAM matrix which defines the size (in number of steps, reported in square brackets) of the adjustments from one stimulus presentation to the next as a function of the participant's answer (yes or no) and of the presence (roll) or absence (no roll) of the motion stimulus.*

Since in the "Roll" condition the thresholds were expected to be the lowest, the initial roll rate was 6 deg/s, with an initial step size of 1 deg/s. For all other conditions the initial rate was 12 deg/s and the step size was 2 deg/s. The step size was halved every 4 reversals of the resulting staircase (Figure 34). The session was terminated after 12 reversals, where a reversal is a change in the staircase direction from decrease to increase or vice versa. The threshold was then computed as the average roll rate over the last 5 reversals. An example of the resulting staircase profile is shown in Figure 34. A typical session lasted about one hour.

After each session, participants filled out a questionnaire to indicate their subjective ratings about mental demand, level of concentration, ability to maintain a constant level of attention, level of frustration, physical comfort and simulation realism on a 9-point rating scale (Table 7).

Simulator Sickness Questionnaires (SSQ) were also collected for all participants after each experimental session to quantify, through the SSQ score, the level of participants' motion sickness (Kennedy et al. 1993). In every session, the level of sickness was monitored every 10 minutes using a numerical score, based on the scale used by Golding et al. (2001). The purpose of this numerical score was to monitor the state of the participants and promptly interrupt/abort the session in case of symptoms of simulator sickness. No statistical analysis was performed on these scores.

6.4 RESULTS

In the following sections we report the results of the three types of measures that were collected during the experiment: perceptual thresholds for roll-rate, objective measures of the driving behaviour, and subjective ratings of the task attentional demand and the level of immersion in the simulation.

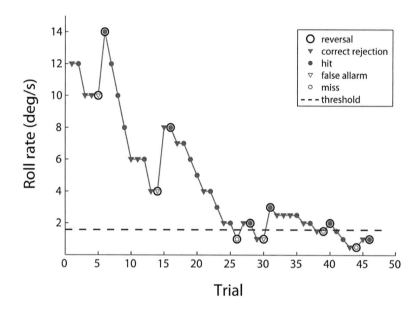

Figure 34 *An example of staircase from the study (condition "+Sway"). The initial roll rate was 12 deg/s, with initial step size 2 deg/s, which was halved to 1 deg/s after 4 reversals and to 0.5 deg/s after 8 reversals. The staircase was terminated after 12 reversals and the last 5 were averaged to compute the threshold (dashed line).*

Item #	Question
1	"Overall mental demand"
2	"Average level of concentration"
3	"Ability to keep concentration"
4	"Level of frustration"
5	"Physical comfort"
6	"Feeling of being in a car"
7	"Quality of lateral motion: strength"
8	"Quality of lateral motion: timing"

Table 7 *Rating Scale.*

One participant did not complete condition "+Visual" due to mild symptoms of motion sickness. Therefore, the corresponding staircase stopped after 7 reversals (31 trials), of which the last 5 were used to calculate the threshold. Two participants did not fill out the questionnaires at the end of a session. The missing values were replaced by the average score of the other participants in the same condition, and included in the analysis of subjective ratings.

Perceptual thresholds

Mean detection threshold for roll-rate increased from 0.7 deg/s with roll only (condition "Roll") to 6.3 deg/s in active driving (condition "+Active"), while mean threshold was 3.9 deg/s and 3.3 deg/s in conditions "+Sway" and "+Visual" respectively (Figure 35, blue line). The ANOVA indicated a significant main effect of the four conditions on the roll-rate detection threshold ($F(3,27) = 5.36$, $p < 0.05$). Post hoc test with Bonferroni adjustment for multiple comparisons revealed a significant difference between condition "Roll" and conditions "+Sway" and "+Visual", which did not differ from each other. For condition "+Active", large differences between participants were observed: for some the threshold did not increase from passive to active driving, while for others a threshold about 3 times higher was measured. A cluster analysis (k-means clustering) of the thresholds distribution in condition "+Active" revealed that thresholds values could be divided into two clusters: "High Threshold Cluster" and "Low Threshold Cluster", respectively indicated by black and red lines in Figure 35.

Behavioural measures

During the trials of condition "+Active", steering wheel commands and car position were continuously recorded. These data were analysed to find evidence of the above reported differences in motion sensitivity.

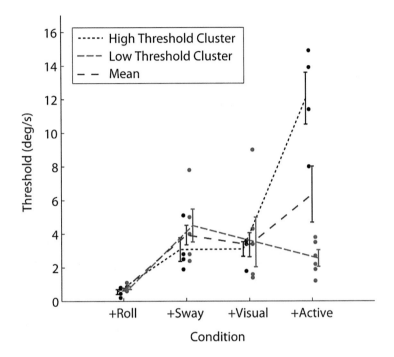

Figure 35 *Roll rate detection thresholds. Global mean (blue line); mean thresholds for participants who showed increment in condition "+Active" (black line); mean thresholds for participants who showed no increment in condition "+Active" (red line). Error bars indicate ±1 SEM.*

Steering wheel commands were averaged for every participant over all trials of the "+Active" condition. Neither the root mean square (rms) of the average trace nor the average variance showed significant correlation with the measured perceptual thresholds. Moreover, they were not significantly different between the two clusters. Similar conclusions were obtained from the analysis of car lateral position on the track, lateral force generated by the MCA in response to participants' commands (i.e. the sway intensity) and number of steering wheel reversals during the curve. An inspection of the power spectral density (PSD) of the steering wheel commands also indicated no qualitative differences in driving behaviour between the two clusters. Overall, the rms of the steering wheel was on average 22 ± 0.4 deg and the rms of the sway was 0.9 ± 0.02 m/s^2. The maximum sway experienced by the participants inside the simulator was 1.1 ± 0.1 m/s^2 and not significantly different between the two clusters.

We did not find any significant differences in driving behaviour for all the considered objective measures. This clearly indicates that the differences in roll-rate detection thresholds between the two clusters cannot be attributed to different driving styles, or to different motion profiles experienced by the participants.

Subjective ratings

In the active driving condition ("+Active"), participants with high thresholds (low sensitivity) reported a lower level of immersion (question 6 "Feeling of being in a car") than participants with low thresholds (better sensitivity), as shown in Figure 36.

The Pearson's correlation coefficient calculated on the rating scores resulted in a significant negative correlation between the "Feeling of being in a car" (question 6) and the roll-rate thresholds ($r = -0.72$, $n = 10$, $p < 0.05$). This indicates that participants who experienced a lower feeling of immersion in the simulation also showed a lower sensitivity for roll rate,

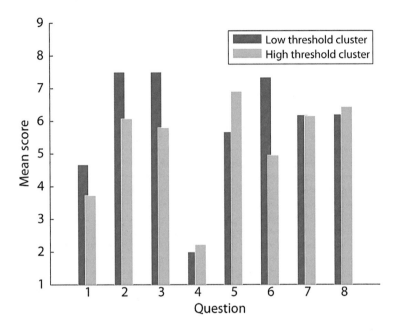

Figure 36 *Mean rating scores from the questionnaires collected at the end of the "+Active" condition reveal a significantly higher level of immersion (question 6) for the participants with lower thresholds compared to those with higher thresholds. This results in a negative correlation between measured threshold values and the subjective level of immersion.*

and were unable to notice the rotation of the platform up to 12 deg/s on average (Figure 35). All other questions did not show any significant correlation with the threshold clusters.

Overall, a significant main effect of the conditions on questionnaire scores emerged from question 1 ("Overall mental demand"; $F_{(3,27)} = 3.67$, $p < 0.05$) and from question 6 ("feeling of being in a car"; $F_{(3,27)} = 6.58$, $p < 0.01$). Post-hoc tests with Bonferroni correction revealed that both the overall mental demand and the feeling of being in a car increased significantly from the condition in which roll motion was presented in darkness ("Roll") to the conditions where active control was available ("+Active"), as shown in Figure 37. Participants' ratings in response to other questions (listed in table 7) were not significantly affected by the experimental condition.

Throughout the experiment simulator sickness was monitored after every experimental condition using the SSQ system introduced by Kennedy et al. (1993). As confirmed by the SSQ scores, mild symptoms of motion sickness were reported by one participant in conditions "+Sway" (SSQ score of 59) and "+Visual" (SSQ score of 97) and required the "+Visual" condition to be prematurely terminated. The SSQ scores of the remaining participants (see Figure 38) were tested by means of a repeated-measures analysis of variance and revealed a significant main effect of the condition ($F_{(3,24)} = 3.99$, $p < 0.05$). Post-hoc analyses with Bonferroni correction were not statistically significant. Note that SSQ scores are overall low, in agreement with participants' verbal reports of no perceived motion sickness. Compared to the SSQ score reported in the aborted condition (97), analysed SSQ scores were on average at least one order of magnitude lower (cf. Figure 38).

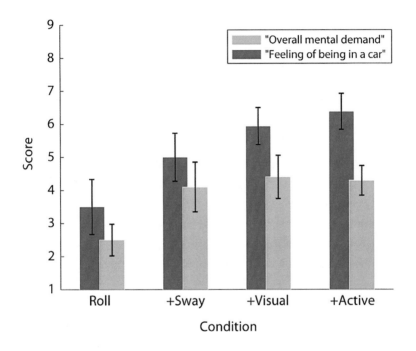

Figure 37 *Mean score at question 1 "Overall mental demand" (orange bars in the front) and at question 6 "Feeling of being in a car" (blue bars in the back) for different conditions. Error bars indicate ±1 SEM.*

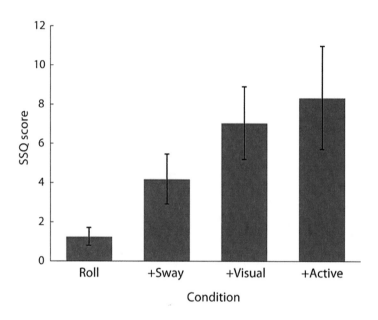

Figure 38 *SSQ scores for 9 participants averaged for the four different conditions.*

6.5 DISCUSSION

The purpose of the study was to investigate the effect of different sensory and cognitive factors on motion sensitivity in a driving simulation. We measured detection thresholds for roll rate in darkness, with additional lateral motion, with available visual motion information, and in conditions that closely resemble an actual driving scenario, in which a driver actively controls the vehicle.

We found that thresholds significantly increase when translational motion (sway) is added to rotational motion (roll). This result essentially replicates what was previously reported by Zaichik et al. (1999). Visually suppressing the roll did not increase the thresholds further, but led values comparable to those reported by Groen and Bles (2004). Our study extends the validity of previous results by allowing a direct comparison of the measured thresholds, since all the conditions were tested here using the same participants, motion stimuli, platform and methodology.

Interestingly, the addition of incongruent visual information did not affect further the detection threshold. Indeed, the visual motion information during the experiment showed a lateral translation with no roll, which was actually present in the inertial motion stimulation due to tilt-coordination. We confirm again the results of previous literature (Groen and Bles 2004), which indicated in about 3 deg/s the pitch rate detection threshold for incongruent visual-inertial stimulation during passive motion.

Up to now, this value was widely adopted within the driving simulation community and used as a reference for the tilt rate limiter in the washout algorithms responsible for tilt-coordination. However, one should consider that this value was measured during passive motion. In other words, this value refers to the motion sensitivity of a passenger, not a driver. Since it is reasonable to expect a further decrease in sensitivity during an active task (Hosman and van der Vaart 1978), we measured the threshold during active driving, with the intent of providing the community with a more realistic value to be used in simulation. The result replicates previous finding by Nesti et al. (2012), indicating an average roll rate detection threshold of approximately 6 deg/s (Nesti et al. 2012). However, here we found that two clusters better describe the thresholds for roll in active driving conditions. The "low threshold" cluster showed no difference in threshold between the passive and active driving simulation (approximately 3 deg/s), while the "high threshold" cluster showed a significant increment: roll rate of 12 deg/s was required to perceive a body rotation to the side while driving.

It is reasonable to hypothesize that differences in drivers' sensitivity emerges when participants are not equally engaged by the "+Active" condition. Indeed, the driving task could be more challenging for a less experienced driver compared to a more experienced one, resulting for instance in more frequent steering wheel adjustments and/or higher mental demand. Collected data provides no evidence relating thresholds measured in the "+Active" condition with typical behavioural measures such as steering wheel rms or control reversals (Cain 2007). However, we found a significant negative correlation between the threshold clusters in condition "+Active" and the feeling of being in an actual car. Thus, lower sensitivity for roll rate correlates with a lower level of immersion in the simulation. Conversely, a better feeling of being in an actual car correlates with a higher sensitivity for rotational rates. We hypothesize that sensitive drivers ("low threshold cluster") take advantage of a better feeling of immersion, and can maintain their sensitivity even when their attention is diverted to the driving task, with complex and rich multisensory stimulation. This suggests that a realistic driving simulation, in which drivers have active control over the vehicle, helps the sensitive drivers to better understand the characteristic vehicle motion. As a consequence, drivers concentrate more on the relevant motion aspects, and maintain a high sensitivity. On the other side, one could be overwhelmed by the richness of the simulation and the effort required in controlling the vehicle. This would prevent the driver to reach a sufficient level of immersion. The consequence would be a distribution of the attentional resources over multiple cues, with a reduced level of attention to the relevant motion aspects, and an increased threshold.

The driving simulation community should carefully evaluate the importance of our findings when transferring the results from simulator studies to production cars. Indeed, recruiting simulator users with low motion sensitivity (high threshold) would increase the perceptual workspace of the simulator, as higher tilt rate saturation values would be accepted. However, our study shows that drivers with low sensitivity also report a lower feeling of immersion, with potential negative impact on the validity of the results for safety and training applications.

Future studies should address more specifically the cause of individual differences in motion sensitivity during active driving simulation. The relationship between subjective feeling of immersion and individual motion sensitivity should be also further investigated. For this, the development of novel and robust method for measuring immersion, perhaps based on perceptual judgments (Wallis and Tichon 2013), would be highly beneficial. This will improve our understanding of human motion perception and the reliability and validity of simulator studies for real world applications.

6.6 CONCLUSION

In this study we investigated how roll rate detection threshold during lateral motion is affected by motion in multiple directions, concurrent visual information, and active control task. Indeed, motion in different directions, multisensory visual-inertial stimulation and vehicle control activities are actual parts of a typical driving simulation. Thus, the question is particularly relevant for the development of efficient motion cueing techniques in driving simulation, in order to ensure the best use of the simulator workspace and provide the user with a realistic driving experience.

The main results indicate that roll rate perception is affected by the combination of different simulator motions. Furthermore, for some drivers an active control task seems to increase detection threshold for roll rate, i.e. impair motion sensitivity; while for others the threshold remain unaffected by the additional attentional load.

We hypothesize that an active control task may induce a better feeling of immersion and a better understanding of the vehicle relevant motion. If this does not occur, however, the overall complexity of the simulation may cause motion sensitivity to decrease.

ACKNOWLEDGMENTS

We gratefully thank Maria Lächele, Reiner Boss, Michael Kerger and Harald Teufel for technical assistance.

GRANTS

AN, PP, SN and ML were supported by funds from the Max Planck Society. This work was also supported by the Brain Korea 21 PLUS Program through the National Research Foundation of Korea funded by the Ministry of Education. The funders had no role in study design, data collection and analysis, decision to publish, or preparation of the manuscript.

BIBLIOGRAPHY

AGARD (1979) Dynamic Characteristics of Flight Simulator Motion Systems. AGARD Advis. Rep. No.144, AGARD, NATO, Neuilly sur Seine, Fr.

Agrawal Y, Bremova T, Kremmyda O, et al (2013) Clinical testing of otolith function: perceptual thresholds and myogenic potentials. J Assoc Res Otolaryngol JARO. doi: 10.1007/s10162-013-0416-x

Ahn H-S, Chen Y, Moore KL (2007) Iterative Learning Control: Brief Survey and Categorization. IEEE Trans Syst Man, Cybern Part C Appl Rev 37:1099–1121. doi: 10.1109/TSMCC.2007.905759

Allum JHJ, Graf W, Dichgans J, Schmidt CL (1976) Visual-Vestibular Interactions in the Vestibular Nuclei of the Goldfish. Exp Brain Res 26:463–485.

Angelaki DE, Gu Y, DeAngelis GC (2009) Multisensory integration: psychophysics, neurophysiology, and computation. Curr Opin Neurobiol 19:452–8. doi: 10.1016/j.conb.2009.06.008

Angelaki DE, Yakusheva TA (2009) How vestibular neurons solve the tilt/translation ambiguity. Comparison of brainstem, cerebellum, and thalamus. Ann N Y Acad Sci 1164:19–28. doi: 10.1111/j.1749-6632.2009.03939.x

Asch SE, Witkin HA (1948a) Studies in space orientation: I. perception of the upright with displaced visual fields. J Exp Psychol 38:325–37.

Asch SE, Witkin HA (1948b) Studies in space orientation. II. Perception of the upright with displaced visual fields and with body tilted. J Exp Psychol 38:455–477.

Baird JC (1997) Sensation and judgment: Complementarity theory of psychophysics. Erlbaum

Baloh R, Honrubia V, Kerber K (2011) Baloh and Honrubia's Clinical Neurophysiology of the Vestibular System, 4th edn. Oxford University Press

Bárány R (1921) Diagnose von Krankheitserscheinungen im Bereiche des Otolithenapparatus. Acta Otolaryngol 2:434–437.

Bárány R (1907) Untersuchungen über den vom Vestibularapparat des Ohres reflektorisch ausgelösten rhythmischen Nystagmus und seine Begleiterscheinungen (Ein Beitrag zur Physiologie und Pathologie des Bogengangapparates). Monatsschr Ohrenheilk 41477–526.526.

Barnett-Cowan M, Dyde RT, Harris LR (2005) Is an internal model of head orientation necessary for oculomotor control? Ann N Y Acad Sci 1039:314–24. doi: 10.1196/annals.1325.030

Barnett-Cowan M, Meilinger T, Vidal M, et al (2012) MPI CyberMotion Simulator: Implementation of a novel motion simulator to investigate path integration in three dimensions. J Vis Exp e3436. doi: 10.3791/3436

Beadnell CM (1924) The psychology of sea sickness. Lancet 203:1289. doi: 10.1016/S0140-6736(01)16769-3

Beghi A, Bruschetta M, Maran F (2012) A real time implementation of MPC based Motion Cueing strategy for driving simulators. 51st IEEE Conf. Decis. Control. Maui, Hawaii, USA, pp 6340–6345

Beierholm UR, Körding K, Shams L, Ma W-J (2008) Comparing Bayesian models for multisensory cue combination without mandatory integration. Adv Neural Inf Process Syst 81 – 88.

Benson A, Hutt E, Brown S (1989) Thresholds for the perception of whole body angular movement about a vertical axis. Aviat Sp Environ Med 205–213.

Benson A, Spencer M, Stott J (1986) Thresholds for the detection of the direction of whole-body, linear movement in the horizontal plane. Aviat Sp Environ Med 1088–96.

Berger DR, Schulte-Pelkum J, Bülthoff HH (2007) Simulating believable forward accelerations on a Stewart motion platform. Max Planck Institute for Biological Cybernetics, Techincal Report No. 159

Berthoz A, Pavard B, Young LR (1975) Perception of linear horizontal self-motion induced by peripheral vision (linearvection) basic characteristics and visual-vestibular interactions. Exp Brain Res 23:471–89.

Bertolini G, Ramat S, Bockisch CJ, et al (2012) Is vestibular self-motion perception controlled by the velocity storage? Insights from patients with chronic degeneration of the vestibulo-cerebellum. PLoS One 7:e36763.

Bertolini G, Ramat S, Laurens J, et al (2011) Velocity storage contribution to vestibular self-motion perception in healthy human subjects. J Neurophysiol 105:209–23. doi: 10.1152/jn.00154.2010

Borah J, Young LR, Curry RE (1988) Optimal estimator model for human spatial orientation. Ann N Y Acad Sci 545:51–73.

Bos JE, Bles W (2002) Theoretical considerations on canal-otolith interaction and an observer model. Biol Cybern 86:191–207.

Brandt T, Dichgans J, König E (1973) Differential Effects of Central Versus Peripheral Vision on Egocentric and Exocentric Motion Perception. Exp Brain Res 16:476–491.

Burnham KP, Anderson DR (2004) Multimodel Inference: Understanding AIC and BIC in Model Selection. Sociol Methods Res 33:261–304. doi: 10.1177/0049124104268644

Butler JS, Campos JL, Bülthoff HH, Smith ST (2011) The role of stereo vision in visual-vestibular integration. Seeing Perceiving 24:453–70. doi: 10.1163/187847511X588070

Butler JS, Smith ST, Campos JL, Bülthoff HH (2010) Bayesian integration of visual and vestibular signals for heading. J Vis 10:1–13. doi: 10.1167/10.11.23.Introduction

Cain B (2007) A Review of the Mental Workload Literature. RTO-TR-HFM-121-Part-II

Carpenter MB, Fabrega H, Glinsmann W (1959) Physiological deficits occurring with lesions of labyrinth and fastigial nuclei. J Neurophysiol 22:222–34.

Cohen LA (1961) Role of eye and neck proprioceptive mechanisms in body orientation and motor coordination. J Neurophysiol 24:1–11.

Crane BT (2012) Fore-aft translation aftereffects. Exp Brain Res 219:477–487. doi: 10.1007/s00221-012-3105-9

Crawford J (1964) Living Without a Balancing Mechanism. Br J Ophthalmol 48:357–60.

Cullen KE (2012) The vestibular system: multimodal integration and encoding of self-motion for motor control. Trends Neurosci 35:185–96. doi: 10.1016/j.tins.2011.12.001

Cutfield NJ, Cousins S, Seemungal BM, et al (2011) Vestibular perceptual thresholds to angular rotation in acute unilateral vestibular paresis and with galvanic stimulation. Ann N Y Acad Sci 1233:256–62. doi: 10.1111/j.1749-6632.2011.06159.x

De Graaf B, Wertheim AH, Bles W (1991) The Aubert-Fleischl paradox does appear in visually induced self-motion. Vision Res 31:845–9.

De Winkel KN, Soyka F, Barnett-Cowan M, et al (2013) Integration of visual and inertial cues in the perception of angular self-motion. Exp Brain Res 231:209–218. doi: 10.1007/s00221-013-3683-1

De Winkel KN, Werkhoven PJ, Groen EL (2010) Integration of visual and inertial cues in perceived heading of self-motion. J Vis 10:1–10. doi: 10.1167/10.12.1

Dichgans J, Brandt T (1973) Optokinetic motion sickness and pseudo-coriolis effects induced by moving visual stimuli. Acta Otolaryng 76:339–348.

Dichgans J, Brandt T (1978) Visual-Vestibular Interaction: Effects on Self-Motion Perception and Postural Control. Handb. Sens. Physiol. Vol. 8 Percept.

Dichgans J, Schmidt CL, Graf W (1973) Visual input improves the speedometer function of the vestibular nuclei in the goldfish. Exp Brain Res 18:319–322.

Doya K, Ishii S, Pouget A, Rao RPN (eds) (2007) Bayesian Brain: probabilistic approaches to neural coding.

Duh HB-L, Parker DE, Philips JO, Furness T a (2004) "Conflicting" motion cues to the visual and vestibular self-motion systems around 0.06 Hz evoke simulator sickness. Hum Factors 46:142–153. doi: 10.1518/hfes.46.1.142.30384

Einstein A (1908) Über das Relativitätsprinzip und die aus demselben gezogenen Folgerungen. Jahrb der Radioakt und Eektronik 4:411–462.

Ellis G (2012) Control System Design Guide, Fourth Edition: Using Your Computer to Understand and Diagnose Feedback Controllers. Butterworth-Heinemann

Ernst MO, Banks MS (2002) Humans integrate visual and haptic information in a statistically optimal fashion. Nature 415:429–33. doi: 10.1038/415429a

Ernst MO, Bülthoff HH (2004) Merging the senses into a robust percept. Trends Cogn Sci 8:162–9. doi: 10.1016/j.tics.2004.02.002

Faisal AA, Selen LPJ, Wolpert DM (2008) Noise in the nervous system. Nat Rev Neurosci 9:292–303. doi: 10.1038/nrn2258

Fechner G (1860) Elemente der Psychophysik. Breitkopf & Härtel, Leipzig

Fernández C, Goldberg JM (1976a) Physiology of peripheral neurons innervating otolith organs of the squirrel monkey. III. Response dynamics. J Neurophysiol 39:996–1008.

Fernández C, Goldberg JM (1976b) Physiology of peripheral neurons innervating otolith organs of the squirrel monkey. II. Directional selectivity and force-response relations. J Neurophysiol 39:985–95.

Fernández C, Goldberg JM (1976c) Physiology of peripheral neurons innervating otolith organs of the squirrel monkey. I. Response to static tilts and to long-duration centrifugal force. J Neurophysiol 39:970–84.

Fetsch CR, Turner AH, DeAngelis GC, Angelaki DE (2009) Dynamic reweighting of visual and vestibular cues during self-motion perception. J Neurosci 29:15601–15612. doi: 10.1523/JNEUROSCI.2574-09.2009

Fisher DL, Rizzo M, Caird JK, Lee JD (eds) (2011) Handbook of driving simulation for engineering, medicine and psychology. CRC Press

Garrett NJI, Best MC (2010) Driving Simulator Motion Cueing Algorithms – A Survey of the State of the Art. 10th Int. Symp. Adv. Veh. Control (AVEC), Loughborough, UK, 22nd-26th August. pp 183–188

Gescheider GA (1988) Psychophysical scaling. Annu Rev Psychol 39:169–200. doi: 10.1146/annurev.ps.39.020188.001125

Gescheider GA (1997) Psychophysics the fundamentals. Lawrence Erlbaum Associates, Mahwah, NJ, US

Gianna C, Heimbrand S, Gresty M (1996) Thresholds for detection of motion direction during passive lateral whole-body acceleration in normal subjects and patients with bilateral loss of labyrinthine function. Brain Res Bull 40:443–7; discussion 448–9.

Gibson JJ (1950) The perception of the visual world. Houghton Mifflin, Boston, MA

Goldberg JM (2000) Afferent diversity and the organization of central vestibular pathways. Exp Brain Res 130:277–97.

Golding JF, Mueller AG, Gresty MAA (2001) Motion sickness maximum around the 0.2 Hz frequency range of horizontal translational oscillation. Aviat. Sp. Environ. Med.

Gong W, Merfeld DM (2002) System Design and Performance of a Unilateral Horizontal Semicircular Canal Prosthesis. IEEE Trans Biomed Eng 49:175–181.

Grant PR, Advani SK, Liu Y, Haycock B (2007) An Iterative Learning Control Algorithm for Simulator Motion System Control. AIAA Model. Simul. Technol. Conf. Exhib.

Grant PR, Artz B, Greenberg J, Cathey L (2001) Motion Characteristics of the VIRTTEX Motion System. 1st Human-Centered Transp. Simul. Conf. Iowa City, Iowa, pp 1–11

Grant W, Cotton R (1991) A model for otolith dynamic response with a viscoelastic gel layer. J Vestib Res 1:139–151.

Green D, Swets J (1966) Signal detection theory and psychophysics, John Wiley. Los Altos, CA

Greig GL (1988) Masking of Motion Cues by Random Motion: Comparison of Human Performance with a Signal Detection Model.

Groen EL, Bles W (2004) How to use body tilt for the simulation of linear self motion. J Vestib Res Equilib Orientat 14:375–85.

Grossman GE, Leigh RJ, Abel LA, et al (1988) Frequency and velocity of rotational head perturbations during locomotion. Exp Brain Res 70:470–476.

Gu Y, Angelaki DE, Deangelis GC (2008) Neural correlates of multisensory cue integration in macaque MSTd. Nat Neurosci 11:1201–10. doi: 10.1038/nn.2191

Guedry FE, Benson AJ (1976) Coriolis cross-coupling effects: disorienting and nauseogenic or not? Nav. Aerosp. Med. Res. Lab. Pensacola, Florida

Guedry FEJ (1974) Psychophysics of Vestibular Sensation. Vestib. Syst. Part 2 Psychophys. Appl. Asp. Gen. Interpret. Handb. Sens. Physiol. Vol. 6. p pp 3–154

Guilford JP (1932) A generalized psychophysical law. Psychol Rev 39:73–85.

Halmagyi GM, Curthoys IS (1988) A clinical sign of canal paresis. Arch Neurol 45:737–739.

Halvorsen WG, Brown DL (1977) Impulse technique for structural frequency response testing. Sound Vib 11:8–21.

Hartmann M, Furrer S, Herzog MH, et al (2013) Self-motion perception training: thresholds improve in the light but not in the dark. Exp Brain Res 231–240. doi: 10.1007/s00221-013-3428-1

Heerspink HM, Berkouwer WR, Stroosma O, et al (2005) Evaluation of Vestibular Thresholds for Motion Detection in the SIMONA Research Simulator. AIAA Model Simul Technol Conf Exhib 1–20.

Henn V, Young LR, Finley C (1974) Vestibular nucleus units in alert monkeys are also influenced by moving visual fields. Brain Res 71:144–149.

Hershenson M (1989) Duration , time constant , and decay of the linear motion aftereffect as a function of inspection duration. Percept Psychophys 45:251–257.

Highstein SM, Fay RR, Popper AN (2004) The Vestibular System.

Hosman R, van der Vaart J (1978) Vestibular models and thresholds of motion perception. Results of tests in a flight simulator. Technical report LR - 265, TU Delft

Jamali M, Carriot J, Chacron MJ, Cullen KE (2013) Strong correlations between sensitivity and variability give rise to constant discrimination thresholds across the otolith afferent population. J Neurosci 33:11302–13. doi: 10.1523/JNEUROSCI.0459-13.2013

Jamali M, Sadeghi SG, Cullen KE (2009) Response of vestibular nerve afferents innervating utricle and saccule during passive and active translations. J Neurophysiol 101:141–149. doi: 10.1152/jn.91066.2008

Johnson WH, Sunahara F a, Landolt JP (1999) Importance of the vestibular system in visually induced nausea and self-vection. J Vestib Res 9:83–87.

Kaernbach C (1990) A single-interval adjustment-matrix (SIAM) procedure for unbiased adaptive testing. J Acoust Soc Am 88:2645–2655.

Kanayama R, Bronstein AM, Gresty MA, et al (1995) Perceptual studies in patients with vestibular neurectomy. Acta Otolaryngol Suppl 520 Pt 2:408–11.

Kandel E, Schwartz J, Jessell T (eds) (2000) Principles of neural science. McGraw-Hill/Appleton and Lange., fourth. McGraw-Hill

Kennedy RS, Lane NE, Kevin S, Lilienthal MG (1993) Simulator Sickness Questionnaire: An Enhanced Method for Quantifying Simulator Sickness. Int J Aviat Psychol 3:203–220.

Kingdom FAA, Prins N (2010) Psychophysics: a practical introduction. Academic Press

Kontsevich LL, Tyler CW (1999) Bayesian adaptive estimation of psychometric slope and threshold. Vision Res 39:2729–37.

Kuo AD (2005) An optimal state estimation model of sensory integration in human postural balance. J Neural Eng 2:S235–49. doi: 10.1088/1741-2560/2/3/S07

Lächele J, Venrooij J, Pretto P, Bülthoff H (2014) Motion Feedback Improves Performance in Teleoperating UAVs. 70th Am. Helicopter Soc. Int. Annu. Forum. Curran, Red Hook, NY, USA, pp 1777–1785

Lackner JR (2014) Motion sickness: more than nausea and vomiting. Exp Brain Res 2493–2510. doi: 10.1007/s00221-014-4008-8

Lackner JR, Graybiel A (1984) Influence of Gravitoinertial Force Level on Apparent Magnitude of Coriolis Cross-Coupled Angular Accelerations and Motion Sickness. NATO-AGARD Aerosp. Med. Panel Symp. Motion Sick. Mech. Predict. Prev. Treat. pp 1–7

Levitt H (1971) Transformed up-down methods in psychoacoustics. J Acoust Soc Am 49:467–477.

Lopez C, Falconer CJ, Mast FW (2013) Being moved by the self and others: influence of empathy on self-motion perception. PLoS One 8:e48293. doi: 10.1371/journal.pone.0048293

Lowenstein O (1955) The effect of galvanic polarization on the impulse discharge from sense endings in the isolated labyrinth of the thornback ray (raja clavata). J Physiol 127:104–117.

Luce RD, Edwards W (1958) The derivation of subjective scales from just noticeable differences. Psychol Rev 65:222–37.

MacGrath BJ, Guedry FE, Oman CM, Rupert AH (1995) Vestibulo-ocular response of human subjects seated in a pivoting support system during 3 Gz centrifuge stimulation. J Vestib Res 5:331–347.

Mach E (1875) Grundlinien der Lehre von den Bewegungsempfindungen. Leipzig: W. Engelmann

MacNeilage PR, Banks MS, DeAngelis GC, Angelaki DE (2010a) Vestibular heading discrimination and sensitivity to linear acceleration in head and world coordinates. J Neurosci 30:9084–94. doi: 10.1523/JNEUROSCI.1304-10.2010

MacNeilage PR, Turner AH, Angelaki DE (2010b) Canal-otolith interactions and detection thresholds of linear and angular components during curved-path self-motion. J Neurophysiol 104:765–73.

Mallery RM, Olomu OU, Uchanski RM, et al (2010) Human discrimination of rotational velocities. Exp Brain Res 204:11–20. doi: 10.1007/s00221-010-2288-1

Maloney LT, Yang JN (2003) Maximum likelihood difference scaling. J Vis 3:573–85. doi: 10:1167/3.8.5

Massot C, Chacron MJ, Cullen KE (2011) Information transmission and detection thresholds in the vestibular nuclei: single neurons vs. population encoding. J Neurophysiol 105:1798–814. doi: 10.1152/jn.00910.2010

Massot C, Schneider AD, Chacron MJ, Cullen KE (2012) The vestibular system implements a linear-nonlinear transformation in order to encode self-motion. PLoS Biol 10:e1001365. doi: 10.1371/journal.pbio.1001365

McConnell KG, Varoto PS (1995) Vibration testing: theory and practice. New York

Meares D, Kaoru W, Scheirer E (1998) Report on the MPEG-2 AAC Stereo Verification Tests.

Melvill Jones G, Young L (1978) Subjective detection of vertical acceleration: a velocity-dependent response? Acta Otolaryngol 85:45–53.

Merfeld DM (2011) Signal detection theory and vestibular thresholds: I. Basic theory and practical considerations. Exp Brain Res 210:389–405.

Merfeld DM, Park S, Gianna-Poulin C, et al (2005a) Vestibular perception and action employ qualitatively different mechanisms. II. VOR and perceptual responses during combined Tilt and Translation. J Neurophysiol 94:199–205. doi: 10.1152/jn.00905.2004

Merfeld DM, Park S, Gianna-Poulin C, et al (2005b) Vestibular perception and action employ qualitatively different mechanisms. I. Frequency response of VOR and perceptual responses during Translation and Tilt. J Neurophysiol 94:186–98. doi: 10.1152/jn.00904.2004

Merfeld DM, Priesol A, Lee D, Lewis RF (2010) Potential solutions to several vestibular challenges facing clinicians. J Vestib Res Equilib Orientat 20:71–7. doi: 10.3233/VES-2010-0347

Merfeld DM, Young LR, Oman CM, Sehlhamer MJ (1993) A multidimensional model of the effect of gravity on the spatial orientation of the monkey. J Vestib Res 3:141–161.

Mergner T, Becker W (1990) Perception of horizontal self-rotations: Multisensory and cognitive aspects. In: Warren R, Wertheim AH (eds) Percept. Control self-motion. Lawrence Erlbaum, Hillsdale London, pp 219–263

Mittelstaedt H (1996) Somatic graviception. Biol Psychol 42:53–74.

Mouchnino L, Blouin J (2013) When standing on a moving support, cutaneous inputs provide sufficient information to plan the anticipatory postural adjustments for gait initiation. PLoS One 8:e55081. doi: 10.1371/journal.pone.0055081

Nahon MA, Reid LD (1990) Simulator Motion-Drive Algorithms: A Designer's Perspective. J Guid Control Dyn 13:356–362.

Naseri A, Grant PR (2011) Difference Thresholds : Measurement and Modeling. AIAA Model. Simul. Technol. Conf. Exhib. pp 1–10

Naseri AR, Grant PR (2012) Human discrimination of translational accelerations. Exp Brain Res 218:455–64. doi: 10.1007/s00221-012-3035-6

Nesti A, Barnett-Cowan M, Macneilage PR, Bülthoff HH (2014a) Human sensitivity to vertical self-motion. Exp Brain Res 232:303–314. doi: 10.1007/s00221-013-3741-8

Nesti A, Beykirch KA, MacNeilage PR, et al (2014b) The Importance of Stimulus Noise Analysis for Self-Motion Studies. PLoS One 9:e94570. doi: 10.1371/journal.pone.0094570

Nesti A, Beykirch KA, Pretto P, Bülthoff HH (2015) Self-motion sensitivity to visual yaw rotations in humans. Exp Brain Res 233:861–899. doi: 10.1007/s00221-014-4161-0

Nesti A, Masone C, Barnett-Cowan M, et al (2012) Roll rate thresholds and perceived realism in driving simulation. Driv. Simul. Conf.

Newman MC, Lawson BD, Rupert AH, McGrath BJ (2012) The Role of Perceptual Modeling in the Understanding of Spatial Disorientation During Flight and Ground-based Simulator Training. AIAA Model. Simul. Technol. Conf. Exhib. AIAA 2012 - 5009

Newman RL (2012) Thirty Years of Airline Loss of Control Mishaps. AIAA Model. Simul. Technol. Conf.

Nieuwenhuizen FM, Bülthoff HH (2013) The MPI CyberMotion Simulator: A Novel Research Platform to Investigate Human Control Behavior. J Comput Sci Eng 7:122–131.

Oman CM (1982) A heuristic mathematical model for the dynamics of sensory conflict and motion sickness. Acta Otolaryngol Suppl 392:1–44.

Paige GD (1994) Senescence of human visual-vestibular interactions: smooth pursuit, optokinetic, and vestibular control of eye movements with aging. Exp Brain Res 98:355–372.

Palmisano S, Chan AYC (2004) Jitter and size effects on vection are immune to experimental instructions and demands. Perception 33:987–1000.

Prsa M, Gale S, Blanke O (2012) Self-motion leads to mandatory cue fusion across sensory modalities. J Neurophysiol 108:2282–91. doi: 10.1152/jn.00439.2012

Pulaski PD, Zee DS, Robinson DA (1981) The behavior of the vestibulo-ocular reflex at high velocities of head rotation. Brain Res 222:159–165.

Reason JT (1969) Motion Sickness - Some Theoretical Considerations. Int J Man Mach Stud 1:21–38.

Richerson SJ, Faulkner LW, Robinson CJ, et al (2003) Acceleration threshold detection during short anterior and posterior perturbations on a translating platform. Gait Posture 18:11–9.

Riecke BE, Schulte-Pelkum J, Avraamides MN, et al (2006) Cognitive factors can influence self-motion perception (vection) in virtual reality. ACM Trans Appl Percept 3:194–216. doi: 10.1145/1166087.1166091

Riecke BE, Schulte-Pelkum J, Caniard F, Bulthoff HH (2005) Towards lean and elegant self-motion simulation in virtual reality. IEEE Proceedings VR 2005 Virtual Reality, 2005 2005:131–138. doi: 10.1109/VR.2005.1492765

Robinson DA (1977) Brain Linear Addition of Optokinetic and Vestibular Signals in the Vestibular Nucleus. Exp Brain Res 30:447–450.

Robinson DA (1981) The use of control systems analysis in the neurophysiology of eye movements. Annu Rev Neurosci 4:463–503.

Robuffo Giordano P, Deusch H, Lächele J, Bülthoff HH (2010a) Visual-Vestibular Feedback for Enhanced Situational Awareness in Teleoperation of UAVs. 66th Am. Helicopter Soc. Int. Annu. Forum. pp 2809–2818

Robuffo Giordano P, Masone C, Tesch J, et al (2010b) A Novel Framework for Closed-Loop Robotic Motion Simulation - Part II: Motion Cueing Design and Experimental Validation. 2010 IEEE Int. Conf. Robot. Autom. pp 3896–3903

Roditi RE, Crane BT (2012) Directional asymmetries and age effects in human self-motion perception. J Assoc Res Otolaryngol JARO 13:381–401. doi: 10.1007/s10162-012-0318-3

Sadeghi SG, Chacron MJ, Taylor MC, Cullen KE (2007) Neural variability, detection thresholds, and information transmission in the vestibular system. J Neurosci 27:771–81. doi: 10.1523/JNEUROSCI.4690-06.2007

Seidman SH (2008) Translational motion perception and vestiboocular responses in the absence of non-inertial cues. Exp Brain Res 184:13–29. doi: 10.1007/s00221-007-1072-3

Seidman SH, Au Yong N, Paige GD (2009) The perception of translational motion: what is vestibular and what is not. Ann N Y Acad Sci 1164:222–8. doi: 10.1111/j.1749-6632.2009.03771.x

Shams L, Beierholm UR (2010) Causal inference in perception. Trends Cogn Sci 14:425–32. doi: 10.1016/j.tics.2010.07.001

Shepherd D, Hautus MJ, Stocks M a, Quek SY (2011) The single interval adjustment matrix (SIAM) yes-no task: an empirical assessment using auditory and gustatory stimuli. Attention, Perception, Psychophys 73:1934–47. doi: 10.3758/s13414-011-0137-3

Solomon JA (2009) The history of dipper functions. Attention, Perception, Psychophys 71:435–443. doi: 10.3758/APP

Soyka F, Giordano PR, Barnett-Cowan M, Bülthoff HH (2012) Modeling direction discrimination thresholds for yaw rotations around an earth-vertical axis for arbitrary motion profiles. Exp Brain Res 220:89–99. doi: 10.1007/s00221-012-3120-x

Soyka F, Robuffo Giordano P, Beykirch KA, Bülthoff HH (2011) Predicting direction detection thresholds for arbitrary translational acceleration profiles in the horizontal plane. Exp Brain Res 209:95–107. doi: 10.1007/s00221-010-2523-9

Stevens SS (1957) On the psychophysical law. Psychol Rev 64:153–181.

Stocker AA, Simoncelli EP (2006) Noise characteristics and prior expectations in human visual speed perception. Nat Neurosci 9:578–85. doi: 10.1038/nn1669

Stratulat AM, Roussarie V, Vercher J, Bourdin C (2011) Does tilt/translation ratio affect perception of deceleration in driving simulators? J Vestib Res 21:1–13. doi: 10.3233/VES-2010-0399

Tahboub K, Mergner T (2007) Biological and engineering approaches to human postural control. Integr Comput Aided Eng 14:15–31.

Tanaka K, Saito H (1989) Analysis of motion of the visual field by direction, expansion/contraction, and rotation cells clustered in the dorsal part of the medial superior temporal area of the macaque monkey. J Neurophysiol 62:626–641.

Tanner T (2008) Generalized adaptive procedure for psychometric measurement. Perception 37:93.

Teghtsoonian R (1971) On the exponents in Stevens' law and the constant in Ekman's law. Psychol Rev 78:71–80.

Telban RJ, Cardullo FM, Kelly LC (2005) Motion Cueing Algorithm Development : Piloted Performance Testing of the Cueing Algorithms. NASA/CR–2005, 213747. 1–183.

Telford L, Howard IP, Ohmi M (1995) Heading judgments during active and passive self-motion. Exp Brain Res 502–510.

Teufel H, Nusseck HG, Beykirch KA, et al (2007) MPI motion simulator: development and analysis of a novel motion simulator. AIAA Model. Simul. Technol. Conf. Exhib. pp 1–11

Valente Pais AR, Mulder M, van Paassen MM, et al (2006) Modeling Human Perceptual Thresholds in Self-Motion Perception. 1–15.

Valko Y, Lewis RF, Priesol AJ, Merfeld DM (2012) Vestibular labyrinth contributions to human whole-body motion discrimination. J Neurosci 32:13537–42. doi: 10.1523/JNEUROSCI.2157-12.2012

Van Asten WNJC, Gielen CCAM, Denier van der Gon JJ (1988) Postural adjustments induced by simulated motion of differently structured environments. Exp Brain Res 73:371–383.

Van Atteveldt NM, Formisano E, Blomert L, Goebel R (2007) The effect of temporal asynchrony on the multisensory integration of letters and speech sounds. Cereb Cortex 17:962–74. doi: 10.1093/cercor/bhl007

Van der Kooij H, Jacobs R, Koopman B, van der Helm F (2001) An adaptive model of sensory integration in a dynamic environment applied to human stance control. Biol Cybern 84:103–15.

Van Drongelen W (2008) Signal Processing for Neuroscientists. Academic Press

Van Wassenhove V, Grant KW, Poeppel D (2007) Temporal window of integration in auditory-visual speech perception. Neuropsychologia 45:598–607. doi: 10.1016/j.neuropsychologia.2006.01.001

Von Helmholtz H (1925) Treatise on Physiological Optics (english translation) translated by J. P. C. Southall. Optical Society of America, Rochester, New York

Vroome AM De, Pais ARV, Pool DM, et al (2009) Identification of Motion Perception Thresholds in Active Control Tasks. 1–20.

Waespe W, Henn V (1977) Neuronal Activity in the Vestibular Nuclei of the Alert Monkey during Vestibular and Optokinetic Stimulation. Exp Brain Res 27:523–538.

Wall C, Merfeld DM, Rauch SD, Black FO (2003) Vestibular prostheses: the engineering and biomedical issues. J Vestib Res Equilib Orientat 12:95–113.

Wall C, Weinberg M (2003) Balance prostheses for postural control. IEEE Eng. Med. Biol. Mag.

Wallis G, Tichon J (2013) Predicting the Efficacy of Simulator-based Training Using a Perceptual Judgment Task Versus Questionnaire-based Measures of Presence. Presence 22:67–85. doi: 10.1162/PRES_a_00135

Walsh EG (1961) Role of the vestibular apparatus in the perception of motion on a parallel swing. J Physiol 155:506–513.

Webb NA, Griffin MJ (2003) Eye movement, vection, and motion sickness with foveal and peripheral vision. Aviat Sp Environ Med 74:622–5.

Weber KP, Aw ST, Todd MJ, et al (2008) Head impulse test in unilateral vestibular loss: vestibulo-ocular reflex and catch-up saccades. Neurology 70:454–463. doi: 10.1212/01.wnl.0000299117.48935.2e

Wei K, Körding K (2009) Relevance of Error: What Drives Motor Adaptation? J Neurophysiol 101:655–664. doi: 10.1152/jn.90545.2008.

Wei K, Stevenson IH, Körding KP (2010) The uncertainty associated with visual flow fields and their influence on postural sway: Weber' s law suffices to explain the nonlinearity of vection. J Vis 10:1–10. doi: 10.1167/10.14.4.

Wei X, Stocker AA (2013) Efficient coding provides a direct link between prior and likelihood in perceptual Bayesian inference. Adv Neural Inf Process Syst 25:1313–1321.

Wertheim AH (1994) Motion perception during self- motion: The direct versus inferential controversy revisited. Behav Brain Sci 293–355.

Wertheim AH, Mesland BS, Bles W (2001) Cognitive suppression of tilt sensations during linear horizontal self-motion in the dark. Perception 30:733–741. doi: 10.1068/p3092

Wichmann FA, Hill NJ (2001) The psychometric function: I. Fitting, sampling, and goodness of fit. Percept Psychophys 63:1293–313.

Widrow B, Glover JR, McCool JM, et al (1975) Adaptive Noise Cancelling: Principles and Applications. Proc IEEE 63:105–112.

Wood SJ, Reschke MF, Sarmiento L a, Clément G (2007) Tilt and translation motion perception during off-vertical axis rotation. Exp Brain Res 182:365–77. doi: 10.1007/s00221-007-0994-0

Yu X-J, Dickman JD, Angelaki DE (2012) Detection thresholds of macaque otolith afferents. J Neurosci 32:8306–16. doi: 10.1523/JNEUROSCI.1067-12.2012

Zaichik L, Rodchenko V, Rufov I, et al (1999) Acceleration perception. AIAA Model. Simul. Technol. Conf. Exhib. pp 512–520

Zanker JM (1995) Does motion perception follow a Weber's law? Perception 24:363–372.

Zupan LH, Merfeld DM, Darlot C (2002) Using sensory weighting to model the influence of canal, otolith and visual cues on spatial orientation and eye movements. Biol Cybern 86:209–230. doi: 10.1007/s00422-001-0290-1